Agricultural Seed Production

Agricultural Seed Production

Raymond A.T. George

Former Senior Lecturer, University of Bath, UK

and

Consultant to the United Nations Food and Agriculture Organization (FAO), Italy

CABI is a trading name of CAB International

CABI Head Office
Nosworthy Way
Wallingford
Oxfordshire OX10 8DE
UK

Tel: +44 (0)1491 832111
Fax: +44 (0)1491 833508
E-mail: cabi@cabi.org
Website: www.cabi.org

CABI North American Office
875 Massachusetts Avenue
7th Floor
Cambridge, MA 02139
USA

Tel: +1 617 395 4056
Fax: +1 617 354 6875
E-mail: cabi-nao@cabi.org

A catalogue record for this book is available from the British Library, London, UK.

Library of Congress Cataloging-in-Publication Data

George, Raymond A. T.
Agricultural seed production / Raymond A.T. George.
p. cm.
Includes bibliographical references and index.
ISBN 978-1-84593-819-2 (alk. paper)
1. Seed technology. I. Title.

SB117.G46 2011
631.5′21--dc23

2011015906

ISBN-13: 978 1 84593 819 2

Commissioning editor: Sarah Hulbert
Editorial assistant: Alexandra Lainsbury
Production editor: Shankari Wilford

Typeset by SPi, Pondicherry, India
Printed and bound by CPI Group (UK) Ltd, Croydon, CR0 4YY.

Contents

Preface

The need to improve and increase food production and supply remains a major global issue, which is frequently referred to as 'Food Security'. The main starting point of the majority of agricultural crops is seed. It can be regarded as the primary agricultural input on which all other inputs, such as labour, irrigation and fertilizers or manures depend for success. This volume concentrates on the production of true seed, which results from botanical seed that has arisen as a result of fertilization or apomixis (as is the case with some of the fodder grasses). Some of the agricultural crops such as *Solanum tuberosum* (the potato) and tropical tuber crops such as cassava and yams are normally produced from propagules such as tubers or stem cuttings which, although frequently referred to as 'seed', are not true seed and have therefore not been included.

This volume has taken the various types of agriculturists and their size of operation into account by drawing attention to the small and subsistence farmers, especially in the tropics and subtropics, with their seed production in addition to the major large-scale seed enterprises.

The book is in two parts and has attempted to include the main agricultural crop species of importance as sources of food. The industrial crops have not been included. Part 1 comprises six chapters concerning the principles and practice of agricultural seed production. In Part 2 the agricultural crop species have been dealt with according to their taxonomy as botanical families, either as single or groups of families per chapter. Some of the larger botanical families such as *Gramineae* and *Leguminosae*, which contain genera of major global importance, have been subdivided, either according to their main method of pollination or as their main uses such as livestock forage or direct human consumption.

A textbook cannot remain up to date with changes and new developments in crop protection chemicals because pesticides are continually being withdrawn and replaced, in some cases their formulation and approved use is modified, and recommended usage is likely to differ from one country to another; their uses and applications have therefore been excluded.

This book has been written with the hope and purpose that it will be used by technical, college and university students during their studies of crop production and agriculture, along with other allied courses and agriculturists, including extension officers who require more seed-oriented information during the course of their professional activities. My purpose and objective has been to humbly but wholeheartedly bring together the necessary information required for the understanding of the seed production of a wide range of agricultural crop species.

I am very grateful to Peter R. Thoday for the very useful discussions and encouragement while I was formulating the scope of the book and for his constructive comments from time to time during its writing. My thanks to Sarah Hulbert, the Commissioning Editor, and Alexandra Lainsbury at CABI for their patience and ever ready advice while I was preparing the manuscript; also Shankari Wilford, Production Editor, for so kindly and efficiently dealing with the production side.

Special thanks are due to our family, Andrew, Christopher, Jane, Joseph and Patrick, who helped with the preparation of figures and took care of the occasional computer mishaps which arose during the manuscript production and who also, from time to time, improved my computer literacy.

Finally my great appreciation to my wife, Audrey, who over many years has not only tolerated my absences on long term stays overseas and shorter missions relating to seed crops but has also been very patient while I have been writing this title.

Raymond A.T. George
Bath
England
March 2011

Acknowledgements

The author is extremely grateful to the following for permissions to reproduce the following photographs and figures:

Figures 3.2, 4.7, 8.2, 8.3 and 12.4, photographs by kind permission of The Real IPM Company (K) Ltd (www.realipm.com), the copyright holders.

Figure 12.3 reproduced by kind permission of Dr J.F. Hancock.

1

From Landrace to Modern Plant Breeding

The Role of Seed in Agriculture and Cultivar Development

The real value of agricultural seed becomes most apparent wherever and whenever there is a shortage of satisfactory seed to sow for the production of the following season's crops. Seed is the starting point for the production of many agricultural cultivated species grown throughout the world and includes cereals, edible oils, pulses, animal and livestock fodders, which are all vital for food security. The higher the standards of seed quality and better the availability of the improved seed, the better will be the opportunities for all farmers, regardless of the size of the areas which each cultivates to produce their crops. In this context, 'seed' refers to true botanical seed, however in some contexts seed may also embrace other planting materials such as plantlets or propagules produced by vegetative propagation such as the tropical tuber crops and potatoes. This volume concentrates on the production of 'true' seed, i.e. seed that has arisen as a direct result of fertilization, or in some species as with some of the grasses by apomixis.

Seed that is being produced for further seed production and multiplication requires more care and attention from the producer than seed that is produced for use as grain or stock feed.

The Early Development of Crop Domestication

The interactions between the evolution of man alongside the further development of crop plants has been discussed by Hancock (2004). Many of man's food crops have been derived from species that started to emerge during the Cretaceous period and were subsequently and gradually domesticated during the last 10,000 years. The gradual domestication has resulted from both environment and selection by farmers and growers:

- Environmental influences:
    - *climate*, including ambient temperature, day-length, soil water status, rainfall;
    - *soil*, nutrient availability, including macro- and micro-nutrients, level(s) of available toxic elements, soil salinity;
    - *effects of pathogens and predators* (macro and micro).
- Farmers' influences:
    - *purposeful or conscious*, saving seed from plants with desirable morphological or other suitable characters or traits;
    - *chance or unconscious*, arbitrary decisions with regard to crop husbandry, including times of sowing, transplanting or harvesting.

The plant materials that have evolved as a result of these influences are usually referred to as *landraces*, *local cultivars* or *ecotypes*; they still remain important sources for ongoing crop production in many developing countries, especially where plant breeding programmes have not yet become the norm. However, they may be regarded as *low input cultivars* (often referred to as 'low technological cultivars'). It is generally recognized that these are well adapted to the local conditions in which they are cultivated.

The landraces remain very valuable sources of germplasm for modern plant breeding. Provision for the recognition and continued use and value of local cultivars and landraces has been made in the *Quality Declared Seed System* (FAO 2006), which ensures that their value and role continues to be recognized.

Cultivar Nomenclature

The term *cultivar* which is used in this volume is as defined in Article 10 of the *International Code for the Nomenclature of Cultivated Plants* (Anon., 1980). It denotes an assemblage of plants that is clearly distinguished by any characters (morphological, physiological, cytological, chemical or others) and which when reproduced (sexually or asexually) retains its distinguishing characters.

This International Code has been accepted by different countries as a basis for legislation relating to plant nomenclature. The terms 'cultivar' and 'variety' are taken as exact equivalents although many national and international Acts and Regulations retain the word 'variety' rather than 'cultivar'. The cultivar is the lowest category under which names are recognized in this code. The term 'cultivar' is derived from *cultivated variety*, or their etymological equivalents in other languages. Article 10 notes that the concept of cultivar is essentially different from the concept of botanical variety, varietas, which is a category below that of species.

All plants in cultivation are clearly named according to these principles. Each has a generic name, a specific name and a cultivar name. For example, if we are referring to a certain cultivated variety (i.e. cultivar) of white clover known to seedsmen and farmers as 'Emerald', it is identified as *Trifolium repens* cv. 'Emerald'. In this example, *Trifolium* is the generic name, *repens* is the specific name (sometimes referred to by taxonomists as the specific epithet) and 'Emerald' is the cultivar name. Quite frequently, in order to ensure the exact species being referred to, we give an 'authority' following the specific or species name. For example, in the case of *Trifolium repens* L., the authority, i.e. the botanist who first described it and named the species was Linnaeus. Thus in this example a standard abbreviation for Linnaeus is added after the species name and it appears in print as L.

An interesting history of the cultivar and cultivated plants as we currently know them has been described and discussed by Thoday (2011).

Subspecies

Different types of the same species, as occur for example in *Beta vulgaris*, are referred to as subspecies, abbreviated to subsp.; for example, the scientific name for fodder beet is *Beta vulgaris* L. subsp. *vulgaris*. The word *taxon* is used for a group at any level in a taxonomic system for the classification of plants; if the taxon is smaller than a subspecies, but includes several groups, each may be referred to as a convarietas, which is abbreviated to convar.; as an example, some workers refer to the different types of *Brassica oleracea* as convars, e.g. cabbage as *Brassica oleracea* convar. *capitata*.

There are some variations in nomenclature, for example within the European Economic Community (EEC), separate species names are still used for field pea (*Pisum arvense* L.) and garden pea (*Pisum sativum* L.) with retention of the original specific epithets given by Linnaeus. But the various types derived from the two species are now considered to be too closely related and all of the cultivated types are referred to as *Pisum sativum* L. *sensu lato* (i.e. 'in the broadest sense').

The International Seed Testing Association (ISTA) has produced a *List of Stabilized Plant Names* (ISTA, 2007).

Cultivar Development, The Role of Plant Breeders

Many modern-day cultivars have arisen as a direct result of the application of plant breeding technology and science. Generally they have been bred for specific environments with desirable characters for fresh market or end products. In many cases they became known as *high input cultivars* because they were bred for better agronomic conditions, including adequate soil nutrients and irrigation with available pest and pathogen control. Although this often remains the case, with many other recent cultivars there is also a strong current trend to develop cultivars with characters such as pest, disease and drought resistance. This has largely arisen from increased awareness of the environment, markets developed for organic products, reduced use of pesticides and concerns for farmers' health.

Formulation of Breeding Policy and its Implementation

Plant-breeding policies and the development of improved cultivars may take place in government institutes, private enterprises or universities; in some breeding programmes there can be economic and technical advantages for two or more organizations to collaborate in a combined programme. An interesting example of this type of partnership is the WEMA (Water Efficient Maize for Africa) project, which has four participating institutions and two funding partners (www.aatf-africa.org). This type of cooperation embraces the use of germplasm and/or breeding materials and testing facilities (e.g. evaluating breeding lines for drought tolerance or resistance, facilities for large-scale field trials and evaluations of plant materials). Interestingly, in addition to producing improved maize cultivars as an end product, a great deal of expertise is passed on to local and other workers, which in turn strengthens the research and agronomic expertise in the region.

Another approach described by Heisey and Brennan (1991) is to use an analytical model to evaluate and determine farmers' requirements in order to decide on appropriate breeding programmes.

Participatory Plant Breeding

There is an increasing recognition for the encouragement and establishment of linkages between agricultural research and farmers in developing countries (Eponou, 1996). The improved liaison between the end user and the researcher (including plant breeders) offers an excellent opportunity for farmers and their dependants to express their views and influence the outcome of breeding programmes and the resulting cultivars.

The term Participatory Plant Breeding (PPB) is used for plant breeding programmes in which the developing lines or plant material, while usually still under the jurisdiction of the breeder, is evaluated on a site, or sites, where the breeding lines are cultivated under local growing conditions. The farmers' findings and opinions carry significant weight regarding the future adoption or continued selection of the breeding lines. PPB has the objective of encouraging local growers and communities to participate in the selection process and for them to have an influence on the development of cultivars for their specific agronomic conditions and product requirements. For example, plant breeders and farmers have jointly contributed to the evaluation of bean cultivars in Rwanda (Anon., 1995). The application of PPB has the possibilities of embracing any one or more of the following product users: growers, consumers, marketers, processors, policy makers, nutritionists and health experts. Where required, specialists in food security or gender issues can also be involved in the selection process. The essential advantages of PPB are that developing breeding lines are evaluated under the farmers' local farming conditions and that all crop usage requirements are taken into account during the breeding material's selection and development process prior to it being made available as a cultivar suitable for its purpose. The concept of taking consumers' nutrition into account when planning plant breeding and other research programmes has been outlined by FAO (2001a).

This system of cultivar development and evaluation is used, where appropriate, by many plant breeders in both public and private organizations. For example, the Center for Agricultural Research in Dry Areas (ICARDA) ensures that the local growers' cultural conditions and requirements are taken into consideration in breeding programmes. PPB may also be applied by private seed companies to develop cultivars for use in organic production systems. The World Bank has published a report on participatory selection, plant breeding and varietal change (Walker, 2006).

Plant Breeders' Rights

Plant breeders, or their employing organizations, invest very heavily in the development of new or improved cultivars. It is therefore especially important that there be a royalty on the sale of registered material so as to recoup development and production costs and to receive income for re-investment in further breeding and crop improvements. Thus the granting of plant breeders' rights to the breeder of a new cultivar provides the breeder, or the breeder's institute, protection from the multiplication or reproduction of the cultivar by other unauthorized persons. Furthermore, the registered material is also protected in all other countries which have adopted the system of plant breeders' rights.

The relevant aspects of measures and associated legislation for the protection of the plant breeder have been discussed by Kelly (1994). The Convention of The International Union for the Protection of New Varieties of Plants (UPOV) has been set up to ensure that plant breeders can have exclusive property rights for their new cultivars. A comprehensive account of UPOV and its functions has been described by Greengrass (1998). It should be noted that UPOV uses 'variety' rather than 'cultivar' in its documents. The organization's main publication for information on the development of plant cultivar protection legislation in the world is *Plant Variety Protection* (the UPOV gazette and newsletter), which is published several times each year. The main criteria for a cultivar to be included for protection are that it is:

- novel, i.e. not been previously marketed in the country in which the rights are being applied for;
- distinct from existing, commonly known cultivars;
- sufficiently homogeneous, stable and new (i.e. it has not been commercialized before certain dates established by reference to the date of application for protection); and
- the plants must be homogeneous.

DUS testing

There are specified tests for the above which are always referred to as *Distinctness, Uniformity and Stability* tests, generally referred to as DUS tests. UPOV publishes details of Test Guidelines for descriptions of individual crops. The descriptors are extremely useful for the compilation of cultivar descriptions and for use when roguing or selecting crops for seed production. Those UPOV guidelines which relate to the crops dealt with in Part 2 of this volume are listed in the Appendix.

The criteria for granting of plant breeders' rights are the evaluation of cultivars for distinctness, uniformity and stability (DUS). This enables controlling authorities to regulate the release of cultivars via national lists and to control plant breeders' rights.

It is recognized that while a particular cultivar is a relatively unique genotype there is firstly the possibility that the material may not be completely homozygous. For example, whereas an F1 hybrid would be expected to be composed of a uniform population when grown, an open-pollinated cultivar will display a degree of difference between individuals in the population. In addition, successive multiplications will put different selection pressures on the cultivar. These selection pressures will depend on a range of factors, including the pollination system of the species (i.e. whether predominantly self- or cross-pollinated), the climate and environment where the multiplication takes place, and the criteria used by the person selecting plants to produce the next seed generation.

Other applications of morphological descriptors

The morphological and other characters which assist in the identification of individual cultivars are of particular interest to the seed producer. The knowledge of a cultivar's observable distinguishing characters enables workers to inspect plants for trueness to type as far as possible before they contribute to the gene pool for seed production, i.e. before the start of anthesis. But because some characters (e.g. flower colour) are only visible at certain stages of the plant's development it is useful to have a record (or cultivar description) which assists in the identification at several different stages, especially those which can be observed before any fertilization of flowers has taken place.

The visual characters, such as leaf shape and flower colour, used for identification are referred to as discontinuous characters and may be readily determined at the appropriate plant stage. Characters in this category should be easily identified without disturbing the plant and as far as practicable they should not be significantly influenced by the environment.

Important factors to consider when deciding on morphological characters to use in cultivar classification include the stability of the characters from one generation to the next; unstable characters are not useful. Another consideration is the extent to which the environment used for testing can as far as is possible be the same each year and similarly at each centre where the plants are being evaluated. Major environmental differences, such as the amount of available water and nutrients, will affect the characters assessed.

The Role of the International Plant Genetic Resources Institute (IPGRI)

IPGRI is active in identifying, describing and promoting national and international germplasm collections of a range of species including the many and diverse agricultural species in the world. This international institute has described details of its strategy, plans and their implementation in a policy document (IPGRI, 1993).

The Institute has five Regional Groups, with collective global coverage. IPGRI's main focus is on the developing countries, but the Regional Groups also establish contacts with institutions in the more developed countries.

Cultivar Trials

Cultivar trials can provide information on where priorities should be placed in relation to national seed production or importation, and are essential in developing countries when deciding on a seed programme and the prioritization for seed production or importation of individual cultivars. Lists can be compiled from results of trials to help farmers, extension workers or advisers, and other interested parties. Kelly (1978) defined three types of list which can result from cultivar trials: each contains 'preferred' cultivars (i.e. cultivars which are better than those not listed) and the lists are of three kinds, i.e. descriptive, recommended and restrictive.

Descriptive cultivar lists

This type of list aims to assist farmers to choose from the available cultivars, and would include those that have been demonstrated and statistically upheld in trials as being useful and acceptable. It would exclude cultivars with major faults within a crop group, such as low yield, frost susceptibility and poor morphological characters of the part of the plant which is of economic importance or having poor seed storage characters. There is an increasing emphasis on nutritive values of the end product, especially with a view to improving the nutrition status of subsistence farmers and their dependants. The 'descriptive' list would enumerate the pros and cons of a particular cultivar so that a grower is able to decide which one is best for his particular requirements. Information provided would include suitability for a specific soil type, season, market outlet, pest and disease resistance. 'Descriptive' lists are widely used in countries with a well-developed agricultural industry and can play a major role in developing countries by drawing attention to suitable seed material which is available and would be expected to make a significant contribution to food security.

Recommended cultivar lists

This type of list is relatively short and sets out to advise farmers as to which cultivars are firmly recommended for specific crops and purposes. The authority providing the information (usually government or a public sector agency) is not confining farmers' choice of cultivar to those on the list but clearly advising on the best material available.

Restrictive cultivar lists

A restrictive cultivar list has the objective of restricting the choice of cultivar to those on the list. Cultivars not on the list are prohibited and are not allowed to be offered for sale or to be grown.

Cultivars should not be restricted without good reason, and it is normal practice for a restrictive list to include more cultivars of each crop species than the recommended list. This allows for variations in availability due to fluctuations of seed yields, including seed crop failure resulting from contingencies beyond the control of the seed producer.

Where several countries in close geographical proximity have trading or other agreements, there is now a tendency for them to compile group lists of cultivars for their community. The lists issued by the EEC are an example of this trend.

Cultivar Identification and Classification

Seed is normally sold by the seed trade and purchased by farmers and growers on the understanding that each seed lot is of a named cultivar. In some situations it is important that the cultivar be verified from a sample taken from a seed lot rather than growing on some of the material to observe morphological characters of the mature plants. There are obvious economic advantages in being able to verify a cultivar from a seed sample; these include the possibility of an immediate answer, saving field space and labour. There have been very significant advances in cultivar 'fingerprinting', especially with the applications of DNA profiling, biochemical techniques and computers for the analysis of images. Although some of the so called 'newer methods' have been developed over a long period, their range and applications have accelerated in recent years with significant advances, especially for the more important agricultural crop species. The developments and their possibilities for cultivar identification have been enumerated and discussed by Cooke and Reeves (1998).

Growth chamber and greenhouse testing

These methods of cultivar identification are usually used for the evaluation of a seed sample's purity and are of particular importance for cultivars entered into certification schemes. Payne (1993a) has described techniques, procedures and conditions which are recognized by ISTA for the conduct of a range of specific crop species. For the crops listed, the characters used for identification are at the seedling and/or young plant stage.

Laboratory testing for determining cultivar susceptibility to specific pathogens

Verification of a cultivar's resistance (or susceptibility) to specific pathogens is often of prime importance when assessing material for trueness to type. ISTA has assisted the clarification of approved techniques by the publication of a handbook on this important topic (Schoen, 1993); standards for testing conditions, inocula and media are described with examples of techniques including bean anthracnose and southern leaf spot or blight of maize.

Chemical methods for classification of cultivars

There are at least five so-called 'rapid chemical identification techniques' described by Payne (1993b). These include phenol tests, fluorescence tests, using chemical bases, chromosome number and other chemical testing methods. However, the majority of these are only indicated for the identification of cultivars of agricultural crop species. The fluorescence test may be used to distinguish fodder peas (*Pisum sativum* var. *arvense*) from edible peas (*Pisum sativum* var. *sativum*). When pea seeds are observed under an ultra-violet lamp (wavelength of approximately 360 nm) the testas of fodder peas fluoresce but those of the edible peas do not. Cooke *et al.* (1985) have described a method using vanillin to test freshly harvested seed of broad beans (*Vicia faba*) for the presence of cultivars with tannins. However, as the authors point out, vanillin response in the seed and flower colour when it is grown on are complementary information as the testa is derived from maternal tissue whereas the flower colour observed from growing on could indicate contamination during flowering resulting from unsatisfactory isolation.

The determination of chromosome number can be used as a 'rapid' test to classify beet seedlings (*Beta vulgaris*) according to their ploidy: as tetraploids (28 chromosomes), triploids (27 chromosomes) and diploids (18 chromosomes). In the procedure described by Payne (1993b), young roots not exceeding 2 cm long are taken from 14-day-old seedlings, fixed and stained in preparation for microscopic examination.

Care must be taken when using the tests as some of the reagents carry health and safety risks (Payne, 1993b). It is essential that workers using the fluorescence test wear eyeglasses specified for protection from ultraviolet light.

Electrophoresis for testing of cultivars

The biochemical technique of electrophoresis is generally well established and its application for cultivar identification using seed or seedlings is recognized (Cooke, 1992). Reviews of the uses of this technique for crop taxonomic purposes have been made by Smith and Smith (1992) and Cooke (1995a).

Electrophoresis is defined by Cooke (1992) as a technique used to separate charged particles under the influence of an electric field. Different types of support media have been used to hold the charged particles during separation, including polyacrylamide (hence PAGE being an abbreviation of polyacrylamide gel electrophoresis). Cooke has listed four principal and distinct kinds of electrophoresis which should be recognized in cultivar testing, i.e. native PAGE (or starch GE), PAGE in the presence of sodium dodecyl sulfate (SDS-PAGE), isoelectric focusing (IEF) and two-dimensional (2D) methods. The techniques used for cultivar identification of wheat, barley, peas, ryegrass and maize have been described by Cooke (1992); Lookhart and Wrigley (1995) have also described techniques for the identification of several food grain cultivars using electrophoretic analysis.

DNA profiling techniques for cultivar identification

The use of DNA profiling techniques and the different approaches of its applications for cultivar identification have been reviewed by Cooke and Reeves (1998). They stated that the profiling techniques can be essentially divided into two technological types, i.e. probe-based and amplification.

The probe-based technology includes restriction fragment length polymorphisms (RFLPs). As most of the more reliable RFLPs rely on radioactively labelled probes and autoradiography the application of the technology will be restrictive. An example of the use of RFLPs is provided by Smith and Smith (1991), who reported on the 'fingerprinting' of maize hybrids and inbred lines.

The application of amplification-based DNA profiling is relatively new for the identification of cultivars. The technique referred to as random amplified polymorphic DNA (RAPD) is widely used in laboratories specializing in the development of systems for cultivar identification; it is very likely to become important in the topic of cultivar verification as it does not involve radioactivity, is relatively quick and can be successfully automated. These techniques have been described by Caetano-Anolles (1996); an example of their application has been the identification of soybean cultivars as reported by Jianhua *et al.* (1996).

While the applications of RFLPs and RAPDs have become significant in the development of methods for distinguishing between cultivars, their limitations have been recognized (Cooke and Reeves, 1998). Thus the so called 'second generation' techniques have been attracting attention for the further development of systems for cultivar identification. These more advanced methods include the use of a sequence-tagged microsatellite system (STMS). Morell *et al.* (1995) have described these DNA profiling techniques for a wide range of ornamental and crop species, including soybean and tomato. Another promising advanced method is the use of the technique known as AFLP (analysis of amplified fragment length polymorphisms) and is described by Vos *et al.* (1995). This method has been used by Tohme *et al.* (1996) for the analysis of gene pools in a wild bean collection.

Image analysis for cultivar identification

The various forms of image analysis provide promising techniques for cultivar identification and classification. These include the use of image analysis (IA) recorded by machine, robot or computer and taken from a range of records such as negative or positive photographic material in addition to live plant material without its destruction. The principles involved in IA have been described by Cooke (1995b,c). Image analysis has also been used to measure the size and shapes of French bean pods (van de Vooren and van der Heijden (1993), while Sapirstein (1995) has described the identification of some grain crop cultivars using digital analysis. Fitzgerald *et al.* (1997) have reported the possible use of image analysis for detecting sib proportion in *Brassica* cultivars. The further development and applications of IA are likely to remove any possible risks arising from the human factor during production of data for storage, retrieval and transmission.

Seed Policies for Countries in Transition

The phrase 'countries in transition' may refer either to countries which are re-emerging as independent following political change or countries recovering from a natural or man-made disaster. Examples include nations which were part of the former USSR where there is an ongoing transition from central control to a market economy. In the case of seed supply this has usually entailed a rebalancing of formal and informal seed sectors. Thus more attention has been given in recent years to recovery programmes for seed industries in such countries. For example, this concept is being applied to states in Eastern Europe and elsewhere (FAO, 2001b).

Restoration of Seed Supply and Re-establishment of Seed Industries

Reasons for failure in seed supply

There can be several reasons for a deficiency in farmers' seed supply; seed shortages may result from one or more causes. In some situations the disaster is sudden, while in others the prime cause may have been developing over a longer period, resulting in a slow deterioration in a supply of satisfactory seed. Examples of these reasons are:

- Natural disasters: e.g. drought, flood, earthquake, serious breakdown in pest or disease controls (e.g. locust invasion), decline in genetic and/or other seed qualities or supply resulting from lack of ongoing seed production support, declining field and soil conditions.
- Man-made disasters: political unrest, civil war, international war, lack of a well designed national seed programme or poor implementation of the national seed programme.

The restoration process

It is emphasized that this section on the restoration of seed supply refers to seed that is intended for sowing, not grain intended for immediate food relief.

Restoration of seed supplies can have several approaches. It is customary for replenishment to concentrate on the main staple or staples normally consumed in the area. These, for example, may be maize, specific grain legumes, rice or sorghum. In some situations propagation material of crop species which are vegetatively propagated, e.g. tropical root tubers such as sweet potato or cassava, will be essential requirements for supply of planting materials. In some instances the replenishment of vegetable seed supply is seen as essential for the affected communities' nutrition as not only essential nutritional supplements to the staples, but also for the relatively quick production of short-term crops (George, 2009). Ideally, any seed being supplied for sowing should be open-pollinated cultivars; these are more likely to lead to farmers subsequently adopting improved cultivars and production of further seed generations of these newer or introduced cultivars. Farmers will not be able to produce satisfactory generations from hybrid cultivars.

Sources and suppliers of seed relief

Seed may be provided as a donation from international aid programmes or non-government organizations (NGOs), usually as a relief operation. Some donor organizations may provide funding for seed procurement from satisfactory sources; in these cases the appropriate international supply agency may use the seed standards outlined in the FAO Quality Declared Seed System (FAO, 2006) to evaluate or stipulate the minimum qualities of seed lots which are to be procured and supplied. Effective and sustainable seed relief approaches and restoration activities have been documented by FAO (1999, 2004).

Methods of distributing seed lots following a declared disaster

There are several possible approaches to alleviation of problems presented by the acute shortage of seed required by farmers following a disaster:

- Seed fairs and voucher systems. In this type of distribution farmers are usually assessed as to their seed requirements and given monetory vouchers which are exchanged at the seed fair for quantities of seed that match the voucher's value. The Catholic Relief Service (CRS) has published a manual for seed-based agricultural recovery based on its relief work and experiences in East Africa (CRS, 2002).
- Encourage reliable seed merchants and/or authentic distributors to re-establish their trading networks.
- Encourage seed production at village level by providing stock seed and other essential materials. These are provided to allow local farmers to start, or re-start, seed production activities, especially providing improved cultivars and those which the local farmers are both familiar with and will be most advantageous in the prevailing circumstances.
- In some areas or communities it is important that due attention is given to women farmers who may well be the crop producers; they should therefore be given fair opportunities to participate in all the activities relating to the distribution.

Corollary to emergency seed supply

Once a stable seed supply has become established and is capable of providing farmers with the seed quality and cultivars that they require, national government resources can be concentrated on the various aspects of seed quality control. In the words of President Obama; 'Aid is not an end in itself. The purpose of foreign assistance must be creating the conditions where it is no longer needed'.

It is important that 'gifted', 'donated' or subsidized seed does not jeopardize the development of a viable commercial seed industry which should become the mainstay of a sustainable seed supply. There are several potential players in the follow-up to emergency seed programmes; the following may well have an important role in assisting in the re-establishment of a satisfactory seed supply at local level:

- Support and encouragement for local and area traders.
- Support for national and local seed producers by assisting with procurement of improved stock seed. Ideally, the stock seed should be in line with the morphological and agronomic requirements of the farmers' required types. This should include time of maturity, preferred culinary and possible secondary uses, such as fodder or thatching.

Storage for contingencies

An increasing number of crop programmes in developing countries are adopting the policy of maintaining stored seed stocks for contingencies. The need to safeguard seed stocks for use immediately following severe shortages of stock seed for further multiplication has been widely recognized. The overall implications and requirements of reserve seed stocks for use after major disasters have been discussed by Kelly and George (1998a).

Seed Certification

There are two important aspects for the verification of a given seed lot: (i) seed quality control up to the time it is offered for sale; and (ii) the control of the seed while it is being sold. Therefore seed certification systems have several components.

The different requirements and standards which are examined separately but which contribute to the composite assessment of a seed lot for certification vary from country to country but generally the following components or principles are included:

1. A national designated authority (NDA), which is appointed by the national government, is responsible for implementing the rules of the scheme on behalf of the government. The NDA acts as the inspectorate of the different parts of the scheme and coordinates the findings of the required tests and observations.
2. Cultivars are accepted into the scheme only when shown to be DUS, and have also been shown to be of significant agronomic value.
3. The breeder, or the institute that bred the original cultivar, is responsible for the maintenance of the cultivar and supply of breeder's stock for further multiplication.
4. Each generation of seed is clearly defined, i.e.:

(i) Pre-basic seed: this is seed material at any generation between the parental material and basic seed.

(ii) Basic seed: this is seed which has been produced by, or under the responsibility of the breeder and is intended for the production of certified seed. It is called 'basic seed' because it is the basis for certified seed and its production is the last stage that the breeder would normally be expected to closely supervise.

(iii) Certified seed: this is the first generation of multiplication from basic seed and is intended for the production of ware crops as distinct from a further seed generation. In some agricultural crops there may be more than one generation between basic and certified seed, in which case the number of generations of multiplication after basic seed is stated, e.g. first or second generation agricultural crops such as cereals. It is usually the multiplication rate of a species which determines whether, or not, further generations can be produced beyond the first multiplication from basic seed; for example, some grain legume species have a relatively low multiplication rate and a second multiplication may be allowed.

(iv) Standard seed: this category contains seed which is declared by the supplier to be true to cultivar and purity, but is outside the certification scheme.

5. The agronomic requirements to be observed in planning and production of the seed are defined for specific crops. These include administrative checks on the recent cropping history of the site, its distance from other specified crops and the number of years since crops of the same or related species have been grown on the site.

6. Tests:

(i) Laboratory and control plot tests of the stock seed used for production of certified seed.

(ii) Inspection of the seed crop production field and observations by the designated authority. These include trueness to type, isolation and freedom from specific weeds and seed-borne diseases.

(iii) Sampling techniques to be employed.

(iv) Tests to be done in the laboratory and on field plots with samples to check the identity and purity of the cultivar.

(v) Tests to be done in the laboratory to determine germination, purity and presence of specific seed-borne pathogens.

The success of a seed certification scheme depends on the collection and compilation of evidence from several facets of seed production and quality control. Furthermore, it also relies on demands for certified seed from farmers. Thus a successful certification scheme depends on the ability of a country to produce seed of high quality in sufficient quantities to meet market requirements. It is the culmination of successful seed programmes that, over several or even many years, have established the required infrastructure of a seed industry. If seed certification schemes in a country aim higher than the industry's current realistic capability, then not only will they fail but their shortcomings or failure will be detrimental to any future development of seed legislation or seed control in that country.

The Quality Declared Seed system

The Quality Declared Seed system (QDS) was introduced by FAO (FAO, 1993) and revised in 2006 (FAO, 2006). The scheme is of particular value in countries where there are insufficient resources or lack of infrastructure for the establishment of fully developed seed monitoring systems, such as the production and marketing of Certified Seed. In recent years the QDS system has become important and valuable for:

1. Seed purchased for emergency relief supplies (i.e. seed for sowing).

2. A reference scheme for these purposes, especially in international seed movement.

3. The scheme can also be applied to assist other potential seed suppliers, e.g. farmers' groups and cooperatives, private farms and NGOs who enter into seed supply activities for seed quality assurance.

The system has been designed, and further modified, to ensure that a country's existing seed quality control resources are used to maximum advantage. The concept of the QDS system is to enable farmers and growers to have access to seed material of a satisfactory standard. The QDS scheme also recognizes the role of extension services in the demonstration of improved seed to the farming community. QDS is not intended to be an alternative to, or in competition with, a more developed seed quality control systems, or duplicate the work of specialist organizations. The documentation of the system refers to varieties rather than cultivars. The scheme recognizes three types of varieties:

1. Varieties developed through conventional breeding technologies.

2. Local varieties that have evolved over a period of time under the particular agroecological conditions of a defined area, i.e. 'landraces' or 'ecotypes'.

3. Varieties developed through alternative plant breeding approaches such as participatory plant breeding.

The required organizational framework for the production of QDS on a government's territory should have the following official organizations established along with appropriate staffing and equipment. These include:

- a Seed Consultative Committee and a Variety Registration System; and
- a seed quality control organization.

A QDS Declaration should be made available to the quality control organization for each seed lot (one after planting the intended seed crop and the second following seed processing). The legal framework of the system stipulates that a participating country should establish a list of cultivars eligible for inclusion and that participating seed producers are registered with the relevant national seed authority. The national authority is responsible for checking at least 10% of seed crops entered in the scheme and at least 10% of 'Quality Declared Seed' offered for sale in the country. *The Quality Declared Seed Technical Guidelines for Standards and Procedures* (FAO, 2006) provides crop-specific sections for some 82 crops, including eight food legumes, seven edible oil crops, 16 forage crops, 19 forage legumes in addition to two industrial crops and 35 vegetables; there is provision for adding further crop species in each of the crop groups. The crop-specific sections outline the requirements and obligations for each seed crop, including facilities and equipment, land requirements, field standards, field inspections and seed quality standards.

Truth-in-labelling

This type of seed quality control does not have minimum standards but relies on the vendor making a statement as to the quality of seed offered for sale. The seed seller is obliged by law to state certain facts regarding the quality. The statements made by the seller for individual seed lots are subject to random checking by the government seed control agency; most truth-in-labelling schemes are based on a 10% sampling. The required standards can vary from country to country, and indeed from crop to crop. The truth-in-labelling concept is a useful way of commencing seed quality control in the early stages of a developing seed industry.

Assistance for Seed Industry Development

There are several sources of assistance for the establishment or strengthening of seed programmes in developing countries. Some of the assistance may be financial. The conditions imposed by the donor or lending organization usually depend on the level of agricultural development in addition to the economic or political stability in the recipient country. The following types of organization may be involved in financial assistance for the seed industry's development in a given country.

National government

The assistance may come from national, federal or local government levels or combinations of any of these potential sources. The roles of government and private enterprise in the supply and distribution of seed have been described and discussed in detail by Kelly (1994).

United Nations agencies

The United Nations (UN) generally finances seed programmes via the United Nations Development Programme (UNDP). Occasionally special funding may be available from the UN Food and Agriculture Organization (FAO). A major contribution in the later part of the 20th century was played by the FAO Seed Improvement and Development Programme (SIDP), described by Feistritzer (1981).

Development banks

Examples of these include the World Bank, The Asian Development Bank and The Arab Development Bank. These organizations usually have their own expert staff or consultants who assess requirements; traditionally the banks have generally made loans at fixed rates of interest with repayment over a relatively long period, although with major global changes going on with financial policies the loan conditions may vary.

Bloc aid

This type of funding can be within political groups of countries or from a group of countries formed for trading and other purposes. Examples of such groups who have provided aid for seed-related projects include the European Economic Community (EEC), the Near East Governments' Co-operative Programme and the Arab Agricultural Development Organization.

Bilateral aid

This is the direct financial and/or technical assistance from one country to another. It is not necessarily between adjacent countries, for example a European country may well provide bilateral aid to an Asian or African country, for example Switzerland to Nepal or Germany to Tanzania.

Bilateral aid is often administered through an official organization of the donor country. Examples of these include the Swedish International Development Agency (SIDA) and the Danish International Development Agency (DANIDA) and GTZ of Germany (Deutsche Gesellschaft für Technische Zusammenarbeit). Other forms of financial assistance or aid may be given by organizations such as the United States Agency for International Development (US AID), the Ford and Rockefeller Foundations.

Non-government organizations

The non-government organizations (NGOs) are very active in developing countries. They usually operate on a charitable basis for relief and development work. Although collectively they cover many areas of the world, very few have an obligation exclusive to seed production or seed industry development. Those which include agronomy in their remit are more usually operating in projects confined to relatively small geographical areas, even if the NGO operates regionally or globally. Typical involvements of NGOs in seed projects have been described by Musopole (1995) and Wilson (1995).

Examples of Organizations Operating Internationally to Assist Seed Trading and Policies

The International Seed Federation (ISF)

The International Seed Federation was formed in 1924 to discuss common problems and interests. At that stage of world seed industry development there was relatively little coordinated international activity. The current mission of ISF (see www.worldseed.org) is:

> to facilitate the international movement of seed, related know-how and technology; to mobilize and represent the seed industry at global level; to inform its members; and to promote the interests and the image of the seed industry. In order to fulfil its mission the ISF promotes strong co-operation among national and regional seed associations. It endeavors to work in partnership with organizations responsible for international treaties, conventions and agreements and those that shape policies that impact the seed industry

In addition to dealing with trading problems with which the seed industry is faced from time to time, ISF advises on the settlement of international disputes according to its own arbitration procedures agreed and accepted by members. ISF has sets of rules for each of the main crop groups. The rules deal with such items or aspects of the international trade as contracts relating to offers and sales, import and export licences and definitions of trading terms. Seed quality is also defined in relation to trade, and the information required to be given with a seed lot is listed. Other aspects included are business requirements such as shipping, insurance and packing agreements, and there is a formula for calculating compensations relating to disputes. The organization sees its own rules as being more realistic and appropriate to its arbitration procedures than international law, which can be time consuming and detrimental to satisfactory settlement within the life of a given seed lot.

The ISF cooperates with other international organizations involved with seed, including the International Seed Testing Association (ISTA) and the Organization for Economic Co-operation and Development (OECD).

The Organization for Economic Co-operation and Development (OECD)

The OECD is an inter-governmental organization involved with economic development, employment and expansion of world trade. It has a Directorate for Food, Agriculture and Fisheries, which administers the OECD Seed Schemes; these schemes include rules and directions for seed moving internationally and also the extension of seed certification. The Scheme for Certification of Seed was first introduced by OECD. The organization regularly publishes a list of cultivars eligible for certification. National seed certification schemes usually follow the OECD scheme exactly, or are otherwise very frequently based on it.

Asia and Pacific Seed Association

The Asia and Pacific Seed Association (APSA) was founded in 1994. The association is composed of a wide range of seed enterprises and seed organizations in the region. It has as its main aim the improved production and trade of agricultural and horticultural seed and planting material in the Asia and Pacific region. APSA is an international, non-profit and non-governmental association with a regional forum for the encouragement of collaboration among seed enterprises. The association's activities include the representation of members' interests to governments, as well as compiling and disseminating information on technical, regulatory and marketing issues. In addition it assists in the organization of training and cultivar testing programmes. The association produces a bi-monthly publication, *Asian Seed*. The aims of APSA have been described by Lemonius (1998).

Further Reading

Bruin, M. (2009) The evolution and contribution of plant breeding to global agriculture. *Proceedings of the World Seed Conference, Rome, 8–10 September 2009.* International Seed Federation (ISF), 1260 Nyon, Switzerland.

Cooke, R.J. and Reeves, J.C. (1998) Cultivar identification: review of new methods. In: Kelly, A.F. and George, R.A.T. (eds) *Encyclopaedia of Seed Production of World Crops.* John Wiley and Sons, Chichester, UK, pp.82–102.

CRS (2002) *Seed vouchers and fairs: a manual for seed-based agricultural recovery.* Catholic Relief Services, developed in collaboration with International Crops Research Institute for the Semi-Arid Tropics and Overseas Development Institute, Nairobi.

FAO (1999) *Restoring Farmers' Seed Systems in Disaster Situations.* FAO Plant Production and Protection Paper 150, FAO, Rome.

FAO (2001a) *Incorporating Nutrition Considerations into Agricultural Research Plans and Programmes.* FAO, Rome.

FAO (2001b) *Seed Policy and Programmes for the Central and Eastern European Countries, Commonwealth of Independent States and other Countries in Transition.* FAO Plant Production and Protection Paper 168, FAO, Rome.

FAO (2004) *Towards Effective and Sustainable Seed Relief Activities.* FAO Plant Production and Protection Paper 181, FAO, Rome.

FAO (2009) *Quality Declared Seed for Vegetatively Propagated Crops.* FAO, Rome.

Hancock, J.F. (2004) The origins of agriculture. In: *Plant Evolution and the Origin of Crop Species*, 2nd edn. CAB International, Wallingford, UK, pp.129–150.

International Seed Federation (ISF) Available at: www.worldseed.org (Accessed on 3 February 2011).

Sperling, L., Remington, T., Haugen, M. and Ngala, S. (eds) (2004) *Addressing Seed Security in Disaster Response: Linking Relief with Development, Overview.* International Center for Tropical Agriculture, Cali, Colombia. Available at: www.ciat.cgair.org/africa/seeds.htm (Accessed on 11 January 2011).

Sperling, L., Negada, S. and Tveteraas, A. (2008) *Moving from emergency seed aid to seed security – linking relief with development.* Workshop organized by The Drylands coordination Group Norway and Caritas Norway, in collaboration with The Norwegian Ministry of Foreign Affairs, DGG Proceedings No. 24. Caritas, Oslo.

Thoday, P.R. (2007) *Two Blades of Grass, the Story of the Cultivation of Plants.* Thoday Associates, Box Hill, Corsham, UK.

Tripp, R. and Rohrbach, D. (2001) Policies for African enterprise development. *Food Policy* 26(2), 147–161.

UPOV (2009) *Proceedings of the Second World Seed Conference, Responding to the challenges of a changing world: The role of new plant varieties and high quality seed in agriculture.* UPOV Publication No. 354 (E), Geneva. Available at: www.worldseedconference.org (Accessed on 26 February 2011).

Van Mele, P., Bently, J.W. and Guei, R.G. (eds) (2011) *African Seed Enterprises.* CAB International, Wallingford, UK.

2

Pollination, Types of Cultivar

Pollination

The morphology of different types of flowers, pollen formation and the process of fertilization has been documented by several authors and the reader is referred to Copeland and McDonald (1995).

The transfer of pollen to the stigma is achieved in the flowering plants by either wind, animals (especially insects) or water. For seed production of agricultural crops the main pollinating agents are wind and insects. Flowers which are wind pollinated are said to be anemophilous, e.g. *Zea mays* and *Beta* spp. Flowers which are insect pollinated are entomophilous, examples include many of the cruciferous crops, e.g. *Brassica* spp. The modes of pollination for individual crop species dealt with in this volume are given in Chapters 7–14.

The Use of Insects to Increase Seed Yield

A very large number of insect species are directly involved in pollination. The two most important insect orders concerned with the pollination of seeds of agricultural crops are the *Hymenoptera*, which includes the ants, bees and wasps, and the *Diptera*, the large order which includes the flies. The level of pollinating insect activity in an entomophilous species grown for a seed crop will have a direct effect on seed yield. In many instances the seed produced relies entirely on natural insect pollinations in addition to the roving honeybees maintained by beekeepers. Many specialized seed producers, especially in the USA, ensure adequate pollinator activity while their crops are in flower by hiring colonies of honeybees from professional beekeepers on a contract basis.

A wide range of pesticides is used in modern farming systems (except in crops produced organically) and many of them are toxic to bees and other beneficial pollinating insects. The protection of honeybees from pesticides has been reviewed by Atkins *et al.* (1977).

Generally, cooperation between the beekeepers and the pest control industry will ensure that only the safest recommended materials are promoted for a specific purpose for the safety of pollinating insects. Supplementary hives should be located on the perimeters of flowering crops and placing them in position 2–3 days after pesticide applications will minimize the danger. Pesticides toxic to bees should not be applied while bees or other important pollinators are foraging. Drift from spraying operations on to colonies or flowering crops should be avoided.

Current problems for hive bees and wild bees

One of the major issues relating to pollination in recent years has been the decline in the security of hive bees and the reduction in bumblebee populations. Beekeepers and biologists have outlined the problems which have been identified with these very important and essential pollinating insects. Attempts have been made to bring the matter to the attention of international organizations and national governments, especially in terms of increased funding for related research and extension.

Hive bee problems

Apiarists have always had to be aware of biotic threats to their bee populations. In recent years new problems have arisen, which currently remain unanswered. The most important of these is usually referred to as colony collapse disorder.

The term colony collapse disorder (CCD) has been given to a hive bee colony from which the population has been radically reduced. The symptoms include absence of adult bees (although the queen is present), a young work force and an absence of dead workers either in or near the hive, implying that the missing bees have died while away from their colony. There are several hypotheses as to the cause or causes, including:

- The Israel acute paralysis virus, pesticides, electromagnetic radiation, mobile telephones, unidentified pests or pathogens, current practices by beekeepers including use of antibiotics in colony maintenance and genetically modified organisms (GMOs). However, it must be stressed that none of these suggestions has yet been proven, it must also be pointed out that apparent CCD has occurred in some countries where no GMOs were grown. There is a lot of speculation; serious and properly funded research is required to identify the cause, or causes, as well as satisfactory controls for CCD. The problem has been reported in the USA, Canada and some areas in Europe. However, it is not clear whether all reported outbreaks have started from exactly the same cause of colony collapse.
- Varroa mite (*Varroa* spp.) is a pest that causes deformation of larvae in the hive and also transmits the bee fungus *Nosema ceranea* between and within bee colonies.

Inference for seed producers

Seed producers rely very heavily on bees for pollination of many seed crops. The presence of CCD in an area may deter the contract hive owners from taking their colonies into affected areas, or a premium may be required to offset their possible losses. This can also be the case with outbreaks of other serious hive bee disorders. The supply of bee colonies by contractors is often called 'migratory beekeeping' because the hive owners follow the crops which benefit from supplementary pollinators through the season; this system is very widely practised in the USA.

Bumblebee decline

The decline of bumblebee populations and also of the number of their species has been widely reported. The greatest reduction in numbers has been observed particularly in western Europe and North America. The main reasons for the decline are thought to be the development of intensive farming practices correlated with areas of high human population (Goulson *et al.*, 2008). The reduction of wildflower hay-meadows, clover leys and use of herbicides has reduced the availability of essential nectar and pollen, which are often the only sources of energy for this group of pollinators. In addition, bumblebees have lost some of their nesting sites following reduction of hedgerows or grassy tussocks in rough grassland, depending on their species (Goulson *et al.*, 2008).

In the UK the Bumblebee Conservation Trust advises farmers and growers on suitable actions to take, as well as techniques to apply for the enhancement of wild bee populations.

Alkali bees

The alkali bee, *Nomia melanderi,* is sustainably managed to pollinate seed crops of lucerne, *Medicago sativa*, across western USA where this bee is a native species. It is the world's only intensively managed ground-nesting bee. The bee is used as an effective pollinator in some 2000–4000 ha of lucerne annually; this area represents approximately 25% of the US production. Its establishment as a successful pollinator has been described by Cane (2008). Alkali bees nest in moist loam on bare and level sites; this bee species does not thrive in wet conditions and succeeds best in warm and dry areas.

Leaf cutter bees

The leaf cutter bee, *Megachile rotundata*, is reputed to be easier to manage than the alkali bee. It uses nests in small spaces in wood or other similar nesting sites. It is encouraged by the construction of potential sites provided by the erection of suitable shelters; these consist of wood with 15 mm drilled holes; a successful alternative is the use of bunches of paper drinking straws or similar materials with similar diameter holes; however, it should be noted that plastic materials are not suitable in humid areas where the holes should in any case be shallower. Peterson *et al.* (1992) have discussed the advantages and management of this important pollinator of lucerne in detail. Losses of these important pollinators from toxic pesticides have been significantly reduced following the adoption of Integrated Pest Management (IPM).

Pollen beetles

Pollen beetles (*Meligethes* spp.) are often responsible for low levels of pollen in the flowers of cruciferous crops (cabbage, kales and related species). These small blue-black beetles overwinter in hedgerows, but in warmer weather migrate to the crucifers. This pest is often not noticed until 'blind' stalks are observed on the seed crop plants, by which time it is too late for an effective control. However, a vigilant check on early opening flowers will reveal the beetles among the stamens. Spraying with appropriate insecticides should be done when the plants are in the early bud stage unless significant populations of their natural predatory enemies, including species of *Ichneumonids* or *Braconids*, are observed to be dealing with the problem. The male parental lines should be checked for the absence of pollen beetle to ensure a satisfactory pollen supply and subsequent seed set when producing hybrid seed parent lines of insect-pollinated crops.

Types of Cultivars

Open-pollinated cultivars

An open-pollinated cultivar is defined as a uniform and stable inter-pollinating population at genetic equilibrium. This type of cultivar is generally heterogeneous and heterozygous with gene frequencies at a constant level of expression. This type of cultivar is generally developed from mass selection by eye of individual plants displaying desirable characters, such as time of maturity and morphological characters of economic value or importance. The seeds produced from the selected plants are bulked to form the next generation. The possible disadvantage of this procedure is that if the number of contributing genotype plants is reduced to a low population it leads to inbreeding depression. However, in some cases progeny testing can be applied over a number of years; this removes those individual plants which are contributing deleterious genes to the overall population.

Hybrid cultivars

There are eight types of hybrid cultivars which have been described by Pickett (1998). It is vital that the advice and instructions of the plant breeder responsible for maintaining individual cultivars be adhered to during seed production of the various types of hybrids.

F1 hybrids

This type of hybrid is now available in a wide range of agricultural crop species. An F1 hybrid is produced by crossing two distinct lines. In practice, each of the two parent lines are the result of inbreeding. As far as commercial seed production is concerned F1 hybrid cultivars are the result of crossing two inbred lines which have been maintained under the control or supervision of plant breeders and which are known to produce a desirable hybrid.

The advantages of F1 hybrid cultivars include uniformity, heterosis (hybrid vigour), earliness, higher yield and resistance to specific pests and pathogens, although these are not always all present in any one cultivar.

Male sterility in hybrid seed production

The production of hybrids in self-fertilizing species is possible when male sterile lines can be found for use as female parents. When this is not possible one of several methods of emasculating the flowers of the female parent is adopted:

- Manual removal of anthers before self-pollination can take place: this method is used commercially for hybrid tomato seed production, but is generally only used in breeding programmes rather than commercial production of agricultural crops. A notable agricultural crop where this technique is used is the mechanical or hand removal of the male inflorescences of maize, generally referred to as de-tasseling.
- A chemical hybridizing agent (CHA): the application of a chemical to cause pollen abortion which results in male sterility. It is essential that the chemical does not result in ovule abortion. This technique has been adopted for some agricultural crop species, including cereals and maize.
- A gameticide which inhibits the production of pollen.

Theoretically all the plants in an F1 hybrid cultivar resemble each other exactly, but because some self-pollination of the female parent used in the cross may take place, some plants which are not F1 hybrids occur and are usually morphologically different from the F1. These off-types in an F1 hybrid are usually the result of accidental self-pollination of the female parent and are generally known as 'sibs'. These are so-called because they are the result of 'sister' or 'brother' plants crossing or selfing with the female line.

In addition to the problems associated with sibs in an F1 hybrid seed lot, there are increased production costs compared with open-pollinated cultivars due to the following factors:

- the development of the initial breeding programme;
- subsequent maintenance of the inbred parents;
- extra land required to allow for male parents, care with sowing, isolation and harvesting;
- high labour input when flowers of the female parent have to be emasculated (especially when hand emasculation is necessary); and
- lower seed yield per unit of land sometimes experienced with production of F1 hybrid cultivars.

Double cross hybrids

The double cross hybrids are produced by crossing two pairs of inbred lines. The two resulting F1s are then crossed to produce a double hybrid. This technique has been used in the production of some *Brassica* and maize cultivars and is considered better than single cross hybrids because the seed yield is generally higher, although the maintenance of the appropriate lines and crosses increases the isolation requirements.

Triple cross hybrids

This type of hybrid was described by Thompson (1964) as a method of overcoming the relatively high costs of maintaining inbred lines and is especially important in the production of some fodder brassicas.

Synthetic hybrids

A synthetic hybrid is produced by the mass pollination of several inbred lines selected for their satisfactory combination ability. Random cross-pollination occurs between the different inbred lines, resulting in a mixture of hybrids. The individual inbred lines to be used for the production of these synthetic hybrids are determined by the breeder. Because of the random crossing which takes place there may be some variation from one season to another when the same parents are used. Cross-pollination is normally assured because of the individual inbred lines' relatively high levels of self-incompatibility. This system has been used for some cultivars of *Brassica* species.

Transgenic Technology and Genetically Modified Organisms

The development of biotechnology and its introduction into the seed industry have introduced a new tool for cultivar production in addition to traditional, or orthodox, plant breeding methods outlined above. Techniques for the transfer of genetic material between otherwise incompatible species or genera have become possible. The techniques used are known as *genetic engineering*. The resulting cultivars arising from application of the technology are referred to as a genetically modified organisms (GMOs); cultivars of this origin are also referred as *transgenics*.

The methods used to produce genetically modified plants have been described and discussed by Cooper and MacLeod (1998), who also outline the regulation of genetically modified crops with particular reference to the EC and the UK. They also outline the regulations in the USA, Canada and Japan. Possible applications of genetic engineering include quality improvement of pest and pathogen resistance, quality, resistance to herbicides, producing male sterility, extension of post-harvest storage potential (including shelf life) and nutritional values of a cultivar.

Much of the earlier work relating to crops was led by international companies, but some countries have released GMO material for multiplication by farmers. The development and potential of GMOs has become a very contentious issue. The reader is referred to the relevant publications in the further reading list at the end of this chapter including the Cartagena Protocol on Biosafety to the Convention on Biological Diversity (Secretariat of the Convention on Biological Diversity, 2004).

Many organizations and individuals in the developed countries, such as in much of Europe, appear to have a strong anti-GMO policy and they do not always engage in the point of view of farmers and their dependants in the developing countries. This volume is not the appropriate forum to join in this ongoing GMO debate but it is worth noting that the potential advantages of transgenics for the poorer farmers could include the following, which are given in a very succinct list of the genetically modified (GM) crops that Africans would like to grow and the benefits that the transgenics could bring (Thompson, 2007):

- insect-resistant African maize cultivars;
- crops resistant to African viruses;
- bio-fortified African crops;
- drought-tolerant crops; and
- maize resistant to the parasitic weed *Striga*.

When one looks at evidence from parallel examples with farming authorities and small farmers in developing countries in other regions of the world a similar picture is painted. One of the ironies of the debate, for or against GMOs, is that many of the potential advantages of appropriate GMOs would match the theoretical requirements of those who promote organic crop production, e.g. pest and/or pathogen resistance in cultivars to eliminate the use of chemical pesticides!

The British Crop Protection Council (BCPC) has compiled a global manual listing the genera and the characters which, to date, have been added to existing species with this technology (BCPC, 2010).

Role of the International Seed Testing Association (ISTA) in evaluating seed lots for GM material

- Produce additional seed testing rules for the detection, identification and quantification of GMO material in conventional seed lots.
- Organize proficiency testing for the detection of GMO material in conventional seed lots.
- Set up a platform for the exchange of information between seed-testing laboratories.
- Additional objectives of the Task Force include identification of stacked genes and publication of performance test results and the availability of documentation relating to testing for specific traits. The work and future programmes of the Task Force have been described by Haldeman (2008).

Further Reading

Bees and pollination

Atkins, E.L., Anderson, L.D., Kellum, D. and Neuman, K.W. (1977) *Protecting Honey Bees from Pesticides.* Leaflet 2883, Division of Agricultural Sciences, University of California, Berkeley, California.

Goulson, B., Lye, D.G. and Darvill, B. (2008) Decline and conservation of bumble bees. *Annual Review of Entomology* 53, 191–208.

Plant breeding and seed production

BCPC (2010) *The GM Crop Manual.* BCPC Publications, Thornton Heath, Surrey, UK.

Bruins, M. (2009) The evolution and contribution of plant breeding to global agriculture. *Proceedings of the World Seed Conference, Rome, 8–10 September 2009.* International Seed Federation (ISF), 1260 Nyon, Switzerland, pp.18–31.

Copping, L.G. (ed.) (2010) *The GM Crop Manual*, 1st edn. CAB International, Wallingford, UK.

Feistritzer, W.P. and Kelly, A.F. (eds) (1987) *Hybrid Production of Selected Cereal, Oil and Vegetable Crops.* FAO, Rome.

Ferry, N. and Gatehouse, A. (eds) (2008) *Environmental Impact of Genetically Modified Crops.* CAB International, Wallingford, UK.

Guimaraes, E. (2009) Building capacity for plant breeding in developing countries. *Proceedings of the World Seed Conference, Rome, 8–10 September 2009.* International Seed Federation (ISF), 1260 Nyon, Switzerland, pp.49–56.

Kanungwe, M.B. (2009) Anticipated demands and challenges to plant breeding and related technologies into the future. *Proceedings of the World Seed Conference, Rome, 8–10 September 2009.* International Seed Federation (ISF), 1260 Nyon, Switzerland, pp.32–39.

Murphy, D. (2010) *Plants, Biotechnology and Agriculture.* CAB International, Wallingford, UK.

Niebur, W.S. (2009) The evolution and contribution of plant breeding to global agriculture. *Proceedings of the World Seed Conference, Rome, 8–10 September 2009.* International Seed Federation (ISF), 1260 Nyon, Switzerland, pp.44–48.

Pickett, A.A. (1998) Genetic quality. In: Kelly, A.F. and George, R.A.T. (eds) *Encyclopaedia of Seed Production of World Crops.* John Wiley and Sons, Chichester, UK, pp.71–79.

Spencer, R., Cross, R. and Lumley, P. (2007) *Plant Names: A Guide to Botanical Nomenclature*, 3rd edn. CABI, Wallingford, UK.

Zehr, U.B. (2009) Effective use of modern biotechnology and molecular breeding and associated methods as breeding tools. *Proceedings of the World Seed Conference, Rome, 8–10 September 2009.* International Seed Federation (ISF), 1260 Nyon, Switzerland, pp. 40–43.

Plant breeding and genetics

Allard, P.B. (1999) *The Principles of Plant Breeding.* Wiley, New York.

Brown, J. and Caliglari, P.D.S. (2008) *An Introduction to Plant Breeding.* Blackwell, Oxford, UK.

Secretariat of the Convention on Biological Diversity (2004) *Global Biosafety – From Concepts to Action: Decisions adopted by the first meeting of the Conference of the Parties to the Convention on Biological Diversity serving as the meeting of the Parties to the Cartagena Protocol on Biosafety.* Montreal, Canada.

Xu, Y. (2010) *Molecular Plant Breeding.* CAB International, Wallingford, UK.

3

Agronomy, Maintenance of Cultivar Purity and Organic Crops

Seed Production Areas

The main seed production areas have become established where climatic factors ensure a relatively suitable environment for the production of satisfactory seed crops. These factors include sowing and growing conditions: sufficient rainfall to ensure complete development and maturation of the seeds, but having a relatively dry summer and autumn with relatively little rain and wind. Such conditions enable seeds which are harvested relatively dry to be further processed as necessary before their safe storage. The lack of inclement weather during the final stages of seed development and ripening is also important from the point of view of disease control on the mother plant as a low relative humidity with minimal rainfall and moderate temperatures minimize the development and dispersal of many pathogens (Gaunt and Liew, 1981); for example, Garry *et al.* (1998) have indicated the importance of growth stage and disease intensity in the effects of Ascochyta blight of peas.

Agroclimate Classification

The suitability of a system for the classification of agroclimatic and soil conditions based on the formulation of the 'FAO Variety Description Forms and Variety Passports' has been described and discussed by Kelly (1994). The system was formulated for specific major cereal crop species (i.e. maize, rice, wheat, millet and sorghum). The four main climatic types are designated 'Tropics', 'Subtropics with summer main rainfall', 'Subtropics with winter main rainfall' and 'Temperate'; each of these types is subdivided into Warm, Cool and Cold. A further subdivision is made depending on annual rainfall, i.e. Arid, Semi-arid, Sub-humid and Humid.

Some of the climatic factors necessary for the biennial seed crops include a relatively mild winter to ensure minimal loss of overwintered plants, although for some species there must be sufficiently low temperatures for vernalization.

Soil Requirements

The soil's physical properties have also played a part in the development of seed production areas as generally only those soils with a relatively high water-holding capacity are most suited although they should not be waterlogged in winter. Winter soil conditions are especially important when biennial mother plants are lifted for selection and replanted, or young plants are transplanted into their final seeding quarters, as with some of the seed crops in *Chenopodiaceae*, the beet family. The nutrient status of soils can be modified by applications of appropriate macronutrients or micronutrients but those soils with a satisfactory cation exchange capacity are most useful and are generally easier and cheaper to manage.

Several classical examples of agricultural seed production areas can be cited, including parts of North America, especially the Pacific North-West, and parts of Australia, although there are many others. The oceanic effect in coastal areas prevents overdrying of unthreshed material in the field and can assist in reducing loss from shattering prior to harvesting. Some areas specialize in specific crop groups, for example, parts of East Africa have been utilized for the production of legume seed crops.

As the agricultural industry develops in each country so does their requirements for seed and planting materials. Hrabovszky (1982) discussed the possible effects of crop production in developing countries on seed development plans. There are already many areas where there have been substantial developments of seed production despite adverse climatic conditions; notable examples include Egypt, India, Israel, Kenya and Mexico. This type of development frequently combines climatic attributes such as high summer temperatures with technical inputs such as irrigation. The need to develop cultivars which are more suitable for the production of organic crops has also stimulated the search for new seed production areas in some areas of South America.

The Seed Producer's Commitment

The farmer or grower who is committed to seed production has to be capable of paying attention to detail, of producing a good crop and also have the ability to follow through all the agronomic and legal requirements for seed production.

The production of successful seed crops requires very careful planning and compilation of records relating to the selected site and the background history of the seed stock being used. Other important details of operations such as sowing, use of pesticides, roguing, harvesting, postharvest processing and seed treatments should each be recorded. The overall aim should normally be to achieve a high quality seed lot coupled with high yield without unnecessary inputs or expenditure.

Planning the Crop

The following topics and items should be considered during the planning stage:

- Previous cropping history of the proposed site.
- Likely presence of highly undesirable or parasitic weeds, e.g. wild oat (*Avena fatua*) in a field proposed for cereals, dodder (*Cuscuta* spp.) in a field proposed for clover or a history of *Striga* where it may be proposed to grow maize or sorghum.
- Isolation from other cross-compatible seed and ware crops.
- A good relationship should be developed with neighbouring farmers and private gardeners in the vicinity. An example of pollen contamination from gardens is from *Brassica* spp. that have been allowed to flower when their cropping has finished.
- Arrangements for harvesting, drying and processing of the seed crop, including booking contract harvesters, equipment and facilities; also liaising with contractors, seed traders or other outlets.
- Appropriate registration with seed control and certification agencies according to national or regional requirements.

Previous Cropping

Rotation of related or similar crops is standard agronomic practice and the main reasons include plant nutrition, maintenance of soil physical conditions and minimizing the risk of soil-borne pests, pathogens and weeds common to individual crop groups. The seed producer has to minimize the risk of plant material or dormant seeds remaining in the soil from previous crops which could cross-pollinate or result in an admixture with the planned seed crop. Therefore information of previous cropping on a given site is an important consideration. The relevant Seed Certification Agency or Authority will stipulate the minimum number of years free of specified crops prior to the seed production of each crop species included in a national certification scheme.

Plants derived from dormant seed or vegetative parts of any previous crops are usually referred to as 'volunteers'. The term 'admixture' usually refers to the addition of seed material that does not conform to the relevant specification for purity of the seed lot in question; it also refers to related genetic material that has been inadvertently mixed in during sowing, harvesting or processing. Therefore attention must be paid to the number of years since a related crop was grown on the same site and the longevity and dormancy of weed seeds or related crops compared with the intended seed crop. These periods should be stringently adhered to for the production of pre-basic, basic and certified seed categories and, in practice, for the production of any seed class, whether or not there is relevant or enforced seed legislation.

The need for organic seed producers to use crop species in *Leguminosae* to increase the nitrogen status of their soils must not be allowed to compromise the requirement for sufficient length of time between related crops which could add to the risk of disease or genetic contamination from dormant seed or volunteer plants. Conversely, the inclusion of legume and/or green manure crops in the rotation can help to increase the time break between unrelated crop species with common seed-borne pests and pathogens. However, it should be emphasized that all the seed used in the rotation for an organic seed production system should also have been produced organically (or in accordance with permitted practice of the prevailing scheme).

The Importance of Weeds and their Effects on Quality of the Harvested Seed Lot

Weed species should be controlled for the following reasons:

- Weeds compete with the crop plants for water, light and nutrients.
- There are some important parasitic weed species, e.g. dodder, broomrape and *Striga* (witch weed), which will impede the crop plant and also produce their own seeds which will reduce the purity of the harvested seed lot, and in turn increase the time and cost of improving the purity of the seed lot due to specific seed processing requirements.
- Weeds present at harvesting will increase the time to complete harvesting, provide impurities in the harvested material and possibly increase the crop's drying time.
- Many weed species are secondary hosts to plant pests and pathogens, including nematodes and viruses.

Site Preparation

One of the most important issues is to ensure that the site does not become contaminated with any other crop or weed seed. All incoming equipment should be cleaned of soil, residual weed or crop seed before entering the designated site. A very efficient age-old technique to minimize the weeds which will compete with the emerging and young crop is the stale seedbed technique

The stale seedbed technique

This is a technique that allows weed seeds to germinate in a prepared seedbed before the crop is sown, thereby significantly reducing the potential weed population which would otherwise compete with the germinating crop. As soon as the first flush of weeds shows the soil is surface tilled to a minimal depth, for example by shallow hoeing. Other alternatives to hoeing include flaming, an application of the herbicide glyphosate or a non-residual contact herbicide. The stale seedbed technique is also referred to as 'the flush system' or 'false seedbed' method of weed control. The seed crop is then sown with the minimum disturbance to the seedbed. This is a long-standing traditional system that was used for many years before the advent of herbicides or organic growing. It is especially useful in fields or plots which are known to have a high content of soil-borne weed seeds.

Production of the Seed Crop

The following points are of special importance in seed production and they should receive special attention:

- The labelling of the stock seed to be sown should be confirmed and recorded.
- All seed sowing equipment such as seed drills should be thoroughly checked for seeds of other crops, especially for remaining seeds of the same crop species. The same precaution applies to all other materials including loose fertilizers or other organic materials, including manures added to the sowing site.
- The isolation of the seed crop should be frequently checked especially prior to sowing or planting and again before the start of anthesis. The isolation from other field crops should be confirmed, and the hedgerows, gateways and headlands should be checked for compatible escape crop species and weeds. Wild and escape plants capable of acting as reservoirs of pests and pathogens should also be removed.

Irrigation

The majority of seed crops in the temperate regions, such as northern Europe, normally rely on rainfall without the addition of irrigation. But in some other areas of the world seed producers use irrigation provided that the water source is available and suitable. Although there are several methods of applying irrigation water to a crop, the two most frequently used application systems are overhead or channels. Overhead irrigation can result in lodging. Channel irrigation reduces the efficient use of harvesting machines such as combines because of the variations in soil surface at the base of the crop.

Seed replacement rate

The seed replacement rate can be defined as 'the proportion of a crop species sown in a given year with improved seed which is either certified or first generation of *improved seed* in comparison with the amount of farmers' own saved seed, or second or third generations sown.' This term is mainly used by workers involved with the planning of a country's seed programme. The seed replacement rate is a very useful indication of the improvement required to increase the amount of improved seed sown by farmers, especially the cross-pollinated species. In this concept the hybrid seed used by farmers is regarded as improved seed.

Seed production villages

A seed village consists of a group of neighbouring and near neighbouring small farmers who are all multiplying the same seed stock of a given cultivar with the common purpose of producing an overall single seed lot for a seed-producing agency or seed procurer. The scheme has been used especially in developing agricultural systems. It is usually organized by, or in liaison with, the national seed authority or its seed certification agency, which would supply the stock seed and also monitor the progress at each site. Farmers may be required to attend a short training course at village level; such courses are aimed at providing farmers with the necessary basic information required for quality seed production. After harvesting, each individual farmer's seed yield is amalgamated to form a blend of lots from the participating farms. Any seed lot from a farm whose seed crop has not been approved is not added to the final blend.

The Maintenance of Cultivar Purity During Seed Production

The maintenance of cultivar purity during seed production is achieved by the combination of several factors, some of which start with crop planning, which includes crop rotation and isolation, while others directly relate to the agronomy and management of the growing crop in the field, including the monitoring of morphological characters according to the cultivar's description and the timely removal of off-types during roguing operations.

Isolation

One major factor during the course of seed production is to ensure that the possibility of cross-pollination between different cross-compatible plots or fields is minimized. This is achieved either by ensuring that those crops that are likely to cross-pollinate are not flowering at the same time (i.e. isolation by time) or that they are isolated by distance.

In addition to the question of undesirable cross-pollination, adequate isolation also assists in avoiding admixture during harvesting and the transmission of pests and pathogens between host crops.

Tolerance limits of genetic contamination

The maximum permissible or acceptable contamination resulting from undesirable cross-pollination will depend on the species and on the class of seed to be produced. It follows that a cross-pollinated crop species will have a higher degree of variation than a self-pollinated crop. While it may be thought desirable to reduce pollen contamination to zero, the amount of permissible contamination will vary with the species and the purpose for which the seed stock is intended (i.e. the intended seed class). Even if it were possible to exclude pollen contamination completely it would not be possible to have a tolerance limit lower than the species' mutation rate (Bateman, 1946).

The higher the class (or category) of seed, the lower the acceptable number of off-types; the tolerance levels for the range of seed classes of a given crop species are specified by seed certification authorities or seed quality control agencies.

Isolation in time

This type of isolation is possible within individual farms or multiplication stations; with careful monitoring it is also possible in 'seed production villages' where only one cultivar of a crop is grown specifically for seed production by cooperating growers or farmers. In the planning stage it is arranged in a way that the cross-compatible crops are grown in successive years or seasons. This principle is easier to achieve in those areas of the world where the climate allows two successive seed crops to be grown in a year. Seed production centres, or organizations, responsible for the multiplication of relatively few cultivars can plan production so that no two cross-compatible cultivars are multiplied simultaneously.

Despite the possibility of isolation in time, there still remains the need to ensure that cross-compatible crops are isolated by distance.

Isolation by distance

This type of isolation is based on the concept that if a seed crop is sufficiently distant from any other cross-compatible crop then the adverse pollen contamination will be negligible. In practice it is impossible to completely prevent 'foreign' compatible pollen reaching a crop, because the wind can carry pollen grains or pollen-carrying insects over relatively long distances.

Regulations or recommendations for isolation distances of specific crops take into account the method of pollination (i.e. whether the species is predominantly self- or cross-pollinated) and the pollen vector (i.e. insect or wind). The specified distances are also greater for classes of seeds to be used for further multiplication than for those intended for distribution to growers for the production of a ware crop. The generally accepted isolation distances for individual crops or generic groups are given in Chapters 7–14 of this volume, however they should only be considered as guidelines because the regulations vary between countries. Much of the experimental work investigating optimal isolation distances has been done in temperate regions and factors such as topography, prevailing wind, insect species and insect populations can influence the effectiveness of isolation distances. In practice there are several possible outside sources of contaminating pollen during anthesis of a seed crop, examples of which include:

- compatible crops in cultivar trials;
- private gardens, although for agricultural crop species this applies mainly to the cruciferous, maize and *Beta* seed crops;
- cross-compatible commercial crops;
- volunteer plants; and
- wild or escape species.

Sources of Contamination from GMOs

Other sources of potential contamination from GMOs during the course of seed production will increase as the number and types of GMOs released for field production increases.

Present examples of these sources of contamination from GMOs include:

- admixture with GMO material during harvesting, seed processing or packaging, usually referred to as 'adventitious' material;
- genetically engineered male lines designed for hybrid production;
- glyphosate-tolerant cultivars or lines;
- cultivars engineered for changes in starch composition, e.g. maize for agricultural outlets;
- cross-compatible materials, especially within *Chenopodiaceae* and *Cruciferae*, which have been developed for other agricultural outlets or bio-fuels; and
- undesirable characters which have entered or contaminated other seed crops, weeds or garden plants with a common period of anthesis.

Zoning

In addition to the primary isolation requirement for a seed crop there are, in some countries, zoning schemes for specified geographical areas, which are intended to control the species to be grown either for market or as seed crops. In principle the specification of what is allowed to be grown in a given zone ensures that cross-compatible species, subspecies or types of related crops do not freely cross-pollinate. For example, there are different types of *Beta* species which are all cross-compatible and largely wind pollinated; they include mangold, fodder beet, sugarbeet, Swiss chard and red beet. By allowing only one of these types to be grown in a specified area or zone the chances of highly undesirable cross-pollination between any combination of them will be greatly reduced. Zoning arrangements may be made by voluntary agreement between seed producers and/or plant breeders or included in seed production legislation within individual countries.

Zoning regulations may call for the registration of any seed or market crop in a specified area regardless of the purpose for which it is being produced; this is especially likely for crop species which flower in the course of the development of the marketable crop. In the USA sweetcorn seed is produced in Idaho where it is isolated from maize with which it freely cross-pollinates (Delouche, 1980).

The concept of zoning could be applied to the production of GMO seed crops, although more care would have to be taken with all compatible plant species in the zone in addition to the seed crops.

Discard-strip technique

According to Dark (1971) the pollen concentration in the air over a field of a wind-pollinated crop increases from the windward edge downwind with a tendency to decline again at the lee edge. Therefore during the period of anthesis when the wind would have been blowing from every direction, a strip around the field's perimeter would have received relatively little of the crop's own pollen and there would have been maximum concentration in the centre. The marginal strip is important in the production of genetically pure seeds. When a cloud of contaminant pollen passes over the field a small number of pollen grains will drop out at random. Those falling over the centre of the plot will compete with the relatively high concentration of the crop's own pollen and have a lower chance of fertilizing, whereas those which fall on marginal areas will not have so much competition and will therefore have a higher chance of fertilization. Thus if the seeds from a 5 m wide strip around the perimeter of the plot are harvested separately, according to the genetic quality found when testing a sample by growing it on, they can either be destroyed or placed in a lower seed category. The bulk of the seed will come from the inner area of the plot and can be kept as a separate seed lot.

Griffiths (1950) reported the effects of isolation distances on inter-varietal crossing of *Lolium perenne* in that contamination decreases rapidly at first with increasing distance, but that there is a progressive reduction in the rate of decrease produced by increases in isolation distance. Other workers have found that with ryegrass the first row of a seed plot sited at 8 m from the seed source of contamination can receive up to 42% of contaminating pollen, while the rows situated at 30 m and 120 m from the source of contaminating pollen had 6.0% and 0.8% contamination, respectively (Anon., 1952).

Most pollen contamination of either wind- or insect-pollinated crops occurs around the perimeter of the plot or field. Therefore if the area of each crop is kept as near as possible to a square then fewer seeds are likely to have been produced as a result of undesirable pollen and the minimum amount of seed is involved if the discard strip is applied.

Cultivar Maintenance

During the course of cultivar maintenance or multiplication to increase the quantity of available seed, it is necessary to ensure that the product will be 'true to type'. This trueness to type infers that the plants grown from the latest seed generation do not differ significantly from the cultivar's description. The loss, or deterioration, of a cultivar's unique characters is sometimes referred to as 'loss of trueness to type' or 'running out'. Trueness to type may be referred to as the 'cultivar purity standard' and certifying authorities clearly state the maximum distinguishable off-types. Only seed stocks conforming to a cultivar description should be certified.

The seed crop is inspected at stages to ensure that any undesirable plant material is removed as far as possible before pollination and subsequent seeding takes place. This process of removing 'off-types' is frequently referred to as 'roguing'. The roguing of seed crops for maintaining genetic purity has been reviewed by Laverack and Turner (1995).

The generally accepted roguing stages for individual species or genera of agricultural seed crops are given in Chapters 7–14 of this volume. However, plant breeders who are responsible for the maintenance of individual cultivars, or certification authorities responsible for specifying crop inspections for certification purposes, will each produce their own criteria for the different stages of seed crops. The level of uniformity is usually higher for the predominantly self-pollinated crops than for the cross-pollinated species.

The intensity of roguing or selection during seed multiplication can have a considerable effect on the genetic quality of successive generations. Dowker *et al.* (1971), investigating the effect of selection on bolting resistance and quality of *Beta* spp., found that the type of selection imposed during multiplication and the season during which selections were made produced seed stocks with bolting differences in subsequent generations.

There are different methods of ensuring that the most desirable plants in a population are used for the maintenance of seed stocks: these are 'positive selection' and 'negative selection'. The selection by either method is done at plant growth and development stages; a most important stage for many crops is the start of anthesis, at which stage flower colour is an important character for some crops, for example the red rice off-type, or rogue type is best identified at this stage but the sooner it is removed while identifiable the least chance it has to contribute to the gene pool of the following generation. This principle of early removal applies to any off-typc in a seed crop.

Positive selection

This system may also be referred to as 'mass selection'. The plants are normally selected to a very high standard while still in their vegetative stage. The percentage of plants selected by this method depends on the observed degree of variation in the crop, but the selection of approximately 10% from the overall population is seldom exceeded. It is most commonly used for maintaining breeders', pre-basic and basic seed stocks. Selected plants are usually initially indicated in the field with a cane and may be evaluated more than once depending when the desirable characters are best observed. In biennial crop species it is customary to lift the plants and replant them in a block. In some cases the selected plants are lifted and transferred to a structure, cage or isolated area. It is important that the selected plants are moved to their isolated flowering quarters before the remaining unselected plants start flowering. If the selected plants are to remain *in situ* then the unselected plants are removed and discarded before they commence flowering. The seeds from a positive selection are harvested in bulk unless it is planned to progeny test the harvested seeds, in which case the seed is collected in separate lots from the individual plants (see Faulkner, 1983). Figure 3.1 illustrates a series of isolation cages each fitted with bee colonies.

Negative selection

This system is frequently referred to as 'roguing'. It is the method most frequently used for open-pollinated seed crops in the final stages of multiplication. Those plants which are observed to be within the limits of the cultivar description are retained for seed production, usually *in situ*, while only those plants which are considered to be atypical of the cultivar are removed. Although the percentage of plants 'rogued' out differs according to the species' method of pollination and purity of the seed stock, the proportion of plants removed may be approximately 20%. Thus if 20% are removed by negative selection, it could also be regarded as a positive selection of 80%. Therefore the system of negative selection provides a much lower

Fig. 3.1. Isolation cages provided with bees for pollination of selected brassica crops.

selection pressure than positive selection and is normally used in the final multiplication stage of an open-pollinated seed crop. The resulting seed crop is bulk harvested.

Suitable pathways are left unsown in some high density crops, such as cereals, to provide access to view the developing seed crop during roguing and inspections; these pathways are sometimes referred to as 'tramways'; they are not normally required for wider spaced row crops.

The range of off-types or rogues identified in a crop produced for seed production may include:

- hybrids, (occurring from out-crossing in the previous generation);
- mutations (depending on the species' mutation rate);
- plants resulting from admixture and/or volunteers; and
- plants which are outside the accepted character range for the cultivar.

Guidelines for efficient roguing

The roguing operation is the practical basis for ensuring that the seed lot which will be harvested will be the highest possible genetic quality. The following points are therefore emphasized in order to achieve an efficient operation:

- Ensure that weed control and crop thinning is done in a timely way so that neither the seed crop plants or weeds obscure the crop.
- Make the roguing inspections as soon as the crop has reached the designated stage, *do not delay*, especially if anthesis is about to start.
- Walk slowly at a rate not exceeding 3km/h.
- Individual members of a roguing gang should be scrutinizing a crop strip of approximately 2m width.
- In bright sunny conditions, as far as is practical, each member of the roguing gang should have the sun behind them; this is especially important where colour is an important character to be observed.
- If the crop is likely to wilt at high temperature the roguing should start early in the day, otherwise important morphological characters may be overlooked.
- Be positioned so that each plant in a row crop is seen individually; or for high population crops such as cereals and grasses ensure that there is no underlap of a worker's view of the crop between members of a roguing gang.
- Remove the entire rogue plant, off-type plant or diseased plant from the field or plot and retain in a front-slung bag and ensure that the rogues are subsequently burnt.
- Do NOT allow rejected plant materials to remain in the seed crop or on the headlands.
- Remove all other cross-compatible plants while roguing.
- Pay particular attention to areas adjacent to plot or field entrances, sites of former collection of plant material, access pathways and places where livestock may have been fed.

Organically Produced Seed

The production of seed for organic farming systems has its own special requirements. Strict adherence to the prevailing specifications for organically produced seed must be adhered to in any country of production for seed to be used for authenticated organic crops.

Background

The demand for organically produced crops for the fresh market became popular with consumers towards the end of the 20th century and has continued to increase in the 21st century. The popularity of organic produce increased especially in some North American, European and southern hemisphere markets, although was not confined to these areas. The call for organic produce has also included various forms of processed or fresh foods. The International Food Policy Research Institute (IFOAM), with its head office in

Bonn, Germany, has a global role with links to member organizations in some 108 countries. IFOAM liaises with the main international seed organizations, including the FAO.

Definition and possible future impact

The term 'organic' crop or crops refers to the method of production. The fundamental principle is that a crop is produced without the use of specified crop protection chemicals or inorganic fertilizers on land that is defined as suitable (i.e. approved by an overseeing organization).

The development of market outlets for organic produce has created a demand for organically produced agricultural seed produced under similar conditions as the organic ware crops. Some governments stated an intention to increase the percentage of crops produced organically, this, in turn, is predicted to increase the demand for organic seed.

Legal framework

As with all official schemes, and with consumer protection in mind, there needs to be an overseeing organization or agency in individual countries to monitor the authenticity of crops. In the European Union (EEC) the main starting point is EC Directive 2092/91, which stipulates that the seed used for organic crop production should also be produced in an organic regime. In the UK, the Soil Association has taken the regulatory role and responsibility for authenticating the seed produced. In other European countries similar organizations have taken on the monitoring tasks. In the USA all producers and handlers of organic crops are required to be approved by a USDA certification agent to be allowed to label or sell a crop as 'organically grown'; the regulations also include standards for seed sources for organic crop production (Bonina and Cantliffe, 2005).

In the early stages of its implementation there were derogations in place which allowed producers of crops destined as organic to use non-organically produced seed (often referred to as 'orthodox seed') if seed of a required cultivar was either not available or unavailable in sufficient quantities; this created planning difficulties for the seed producer.

Organic seed and seed suppliers

Initially seed producers and seed suppliers need to know if there is a market for their potential products. There are several factors to be considered in the case of organic seed. These can include the following:

- Which cultivars to include in seed production programmes, as this will very much depend on the cultivars opted for by the organic producers of ware agricultural crops which relates to customer preference. It should be recognized that some of the ware crops such as cereals will be milled for the production of 'organically' produced flour as also will be some of the grain legumes. The range of agricultural organic products is currently quite significant in many of the market outlets in the developed countries.
- Generally, the agricultural producers will choose cultivars according to market requirements, e.g. suitability for milling, brewing, food processing, fresh produce or pre-packaging. For example, there is a need for millers and bread makers to be fully aware of the characters of some wheat cultivars for the production of improved organic bread in the UK. In view of this requirement the Cereals and Oilseeds Division of the Agricultural and Horticultural Development Board conducted agronomic field trials linked with both milling and baking of selected wheat cultivars (Stanley and Wilcockson, 2010).
- The farmers' organic production regimes, including which cultivars have better pest and pathogen tolerance or control when grown in organic regimes.
- The estimated quantities of seed needed for the required cultivars.
- The commercial risks in opting for production of organic seed.

- The feasibility of producing organic seed with only allowable materials, especially for the biennial crop species, which have a longer exposure period to pests and pathogens in the field.
- A suitable plan for control of pests, pathogens and weeds will have to be made, which will satisfy the requirements stipulated by the monitoring authority; this will require the application of biological controls and integrated control systems, usually referred to as Integrated Pest Management (IPM).
- Additional considerations by the seed producer may include: likelihood of a derogation of a cultivar by the controlling authority and therefore possible loss of sales potential if the seed producer has speculated on the production of such a cultivar.
- Some seed producers have already successfully incorporated organic seed production programmes into their businesses, while others will need to develop acceptable protocols when they are planning to.
- It is generally accepted that the production costs of organic seed are higher than for orthodox seed; this is generally due to lower yields in some species. It has therefore become acceptable to the public who prefer to purchase organic products that they will have a higher retail price.
- Many of the pre-sowing treatments to control, or partly control, pre- and post-emergence pests and pathogens in orthodox seed production are not allowable for organic seed. The later section in this chapter includes pre-sowing treatments, some of which are long standing and also newer techniques allowed for organic seed; some of those listed which include chemical treatment would not normally be allowable.
- Organic seed is subject to the same seed quality regulations as orthodox seed, in addition to the prerequisites stipulated for its authentic production.
- It is essential that the various monitoring authorities allow seed producers the same derogations as the producers of organic crops.

Research and discussions related directly to the needs of organic seed production

It has been generally agreed that there is need for further research and development into the requirements of organic seed production (IFOAM, ISF and FAO, 2001a).

The Conference Report (IFOAM, ISF and FAO, 2001b) stated that two legitimate realities exist, i.e.:

1. Large-scale farmers aiming to supply supermarkets in their home countries and/or export markets where there is heavy reliance on modern cultivars, including F1 hybrids.
2. Farmers' groups in developing and developed countries where local markets are supplied often using 'local varieties' using community-based seed production systems and in some instances involved in participatory plant breeding.

Each of these groups has some different requirements regarding their objectives. Suggestions for further research and development can be identified. These include:

- Identification of new seed production areas that can be shown to have a reduced incidence of those pests and/ or pathogens which are difficult to control in traditional seed production areas, especially in the biennial seed crops.
- Evaluate existing cultivars to determine their suitability for crop production under organic production protocols.
- Identify plant-breeding opportunities for the development of cultivars that fulfil the needs of organic crop production protocols.
- Participatory plant breeding can offer an opportunity for close liaison between breeders and organic growers for the further development of cultivars most suited for organic production.
- The biological control systems can be used to great advantage in some situations during seed production. Figure 3.2 shows a sticky thrip trap; this method traps thrips, which are well known virus vectors in addition to damaging crops.

Fig. 3.2. Use of a vertical sticky sheet to trap thrips in Kenya. Photograph by kind permission of The Real IPM Company (K) Ltd (www.realipm.com).

Nutrition and guidance on the production of organic seed

Crop rotation and approved organic and mineral materials are a part of the concept of organic production. However, from the point of view of seed production care must be taken to ensure that undesirable crop or weed seeds and propagules are not introduced with uncomposted organic materials. Generally, synthetic crop protection materials should not be used. The control of pathogens, pests and weeds has to be achieved by appropriate husbandry (including choice of cultivar, rotation, methods of encouraging natural predators and timely cultivations). With the advent of organic farming systems and increased demands for authentically produced organic crops and products there are various monitoring and advisory agencies. For example, in the UK further information may be obtained from the Soil Association Producer Services and the UK Register of Organic Food Standards (UKROFS). The National Institute for Agricultural Botany (NIAB) has included cultivar trials on UKROFS approved organic farms and published lists of cultivars suitable for organic production.

Pest and Disease Control

The main seed-borne pests and pathogens

The main seed-borne pathogens with the common names of the diseases they cause are given in Chapters 7–14 for the seed crops discussed. It must be emphasized that many of the pathogens listed can be transmitted by other vectors. The nomenclature of the pathogens is in accordance with the Annotated List of Seed-borne Diseases published by the International Seed Testing Association (Richardson, 1990). The Annotated List includes fungi, bacteria, viruses and nematodes which affect seed or seed assessment. Common names have also been listed, but these are likely to vary from one area to another, and some pests and diseases may have more than one common name.

Pre-sowing Seed Treatments

There are several pre-sowing treatments which are used for agricultural seeds, including the application of pesticides for the control of seed- or soil-borne pathogens, modification of seed shape or size and pre-germination before sowing.

Seed treatments for the control of pathogens

The available treatments range from the application of chemicals to the seed as dusts or slurries, to the application of heat via hot water, dry heat or steam–air mixtures. Developments have included the use of antibiotic solutions and more recently the addition of chemicals to hot water.

The need for organic seed treatments to replace the use of disallowed chemical treatments, which have become the norm with orthodox seeds, has been indicated by Groot *et al.* (2004), who have cited the possible use of essential oils and organic acids. The evaluation of hot, humid air seed treatments and fluidized beds have been described by Forsberg *et al.* (2000).

The methods for the application of pesticides to a wide range of seeds have been described by the British Crop Protection Council (BCPC, 2009). A description of the main equipment for treating seeds has been described by Jeffs (1986). Seed fumigation of seed lots to control seed-borne nematode species by methyl bromide fumigation has been replaced by other less hazardous methods. The various seed technologies which are emerging have been reviewed by Van der Burg (2009).

Pelleting and protective seed treatments

Pelleting

Pelleting facilitates the manual and mechanical handling of seeds that are either small or awkwardly shaped. Precision drilling direct in the field or sowing in soil blocks using pelleted seed prior to transplanting can ensure a high degree of accuracy regarding seed placement. Individual seeds are encased in an inert material such as montmorillonite clay. The implications of seed pelleting with special reference to seed quality was reviewed by Tonkin (1979). The pelleting material can also be used to incorporate pesticides or form a coating for seed dressings. The technology for pelleting was gained largely from developments with pelleting sugarbeet seed (Longden, 1975). The pelleting of seed is usually done by specialist companies using proprietary processes at the request of individual seed companies. The development of commercial seed treatments has been discussed by Halmer (1994).

Coating

Seed coating, or film coating, is a technique by which additives such as pesticides, nutrients or nitrifying bacteria can be applied to the seed's external surface (i.e. the testa). But in contrast to pelleting the coating conforms to the seed's shape and does not normally significantly modify the seed's size. An air-suspension technique originally used in the pharmaceutical industry is used (Wurster, 1959). A film-forming polymer containing the required active ingredient is sprayed on to the seeds while they are suspended in a column of air, which is either heated or unheated. A colouring agent is usually incorporated at the same time and the coating materials dry very quickly, resulting in free-flowing coated seeds. Scott (1989) has discussed the effects of seed coatings and treatments on plant establishment.

Organic seed production and pre-sowing treatments which include pesticides

Many of the very successful seed treatments developed over the years prior to the start of organic production have to be evaluated. For example, the thiram treatment developed by Maude and Keyworth (1967) was widely adopted but cannot be used in the processing of seed intended for the organic market. Traditional treatments such as hot water treatment (HWT) have come back into wider use as also have other allowable methods; in addition there are some interesting new developments which could also be effective for the orthodox crop producers. The present and ongoing challenges for the seed-coating and pelleting industries have been reviewed by Legro (2004).

Postharvest and pre-sowing treatments for organic seed

There is a need for organic seed producers and suppliers to consider availability and application of appropriate postharvest treatments requested by purchasers which are allowed by the controlling authority in the country where the seed will be marketed or used.

The minimum requirements for seed quality, as required by seed legislation, remain the same for all seed regardless of production protocol. However, organic growers may well require more emphasis on monitoring seed for presence of GMO material, from both admixture and genetic contamination.

Hydration treatments

There are two main types of hydration treatment for seed, i.e. priming and moisturization.

Seed priming

The technique of seed priming is in two basic stages and was first described by Heydecker (1978) and Heydecker and Gibbins (1978). The technique has been further developed and several materials have been used for the preparation of the priming solution (Khan, 1992). Polyethylene glycol (PEG) has been the main material used to control seed hydration levels in the priming process.

Rowse (1996a) has described the concept of 'drum' priming. The commercially adopted system (Rowse, 1996b) involves the hydration of seeds to a predetermined water content over a 1–2 day period by placing them inside a rotating drum into which water vapour is released. The drum is mounted on an electronic balance linked to a computer that monitors the seeds' water content and controls water vapour production so that the seeds do not become visibly wet. The seeds are then incubated in a rotating drum for a period of up to 2 weeks before being dried to the moisture content suitable for their safe storage. The drum priming method avoids the problem of PEG disposal which occurred with priming commercial seed quantities. Large-scale seed priming techniques and their integration with crop protection have been discussed by Gray (1994).

Combining hydration and coating

Taylor *et al.* (1991) have described a system for upgrading *Brassica* species, which combines seed hydration and coating. The seed is first hydrated in water or primed; during this process sinapine leakage commences from the non-viable seeds, the seeds are then coated with an adsorbent which retains the leakage on individual seeds. Following drying, the seeds are exposed to UV light and sorted according to their fluorescent coatings. The non-fluorescent fraction produced an increased proportion of normal seedlings than the control, while the fluorescent fraction contained a higher proportion of dead seeds, or seeds which subsequently produced abnormal seedlings. Taylor *et al.* (1991) have described a sorting method for *Brassica* spp., using a colour sorter with a UV light.

Fluid drilling

The term 'fluid drilling' is used when seeds which have been pre-germinated are sown. Initially it was of particular interest to the vegetable production industries; the techniques and prospects for fluid drilling of vegetable crops were reviewed by Salter (1978a,b). Agronomists involved with some of the agricultural crops have also started to adapt the technology for grasses and some other crops. Hardegree and Van Vactor (2000) investigated the seed priming of *Elymus lanceolatus* (thickspike wheatgrass), *Pseudoroegneria spicata* (bluebunch wheatgrass), *Poa sandbergii* (Sandberg bluegrass) and *Elymus elymoides* (bottlebrush squirrel tail). These grass species are cool-season range grasses that have to compete with wild grass species and other annual weeds for moisture during germination and emergence. The authors reported favourable results for seed priming especially for germination and emergence at the earlier, cooler sowing dates.

Further Reading

History and development

Thoday, P. (2007) *Two Blades of Grass, the Story of the Cultivation of Plants.* Thoday Associates, Box Hill, Corsham, UK.

Progeny testing

Faulkner, G.J. (1983) *Maintenance, Testing and Seed Production of Vegetable Stocks.* Vegetable Research Trust, NVRS, Wellesbourne, Warwick, UK.

Cultivar purity

Haskell, G. (1943) Spatial isolation of seed crops. *Nature* 153, 591–592.

Integrated Pest Management (IPM)

Bailey, A., Grant, W.P., Chandler, D., Greaves, J., Prince, G. and Tatchell, M. (2010) *Biopesticides: Pest Management and Regulation.* CAB International, Wallingford, UK.

BCPC (2010) *The BCPC Pesticide Manual*, 15th edn. BCPC Publications, Thornton Heath, Surrey, UK.

Plant breeding

Bruins, M. (2009) The evolution and contribution of plant breeding to global agriculture. *Proceedings of the World Seed Conference, Rome, 8–10 September 2009.* International Seed Federation (ISF), 1260 Nyon, Switzerland.

Van der Burg, J. (2009) Raising seed quality: what is in the pipeline? *Proceedings of the World Seed Conference, Rome, 8–10 September 2009.* International Seed Federation (ISF), 1260 Nyon, Switzerland, pp.177–185.

4

Harvesting and Processing

Successive Seed Development

The duration of anthesis of some crop species is prolonged, especially in those species which have successive flowers or complex inflorescences resulting in a prolonged period of successive flowering, seed development and maturation. In many, e.g. *Brassica* species, sesame and clover there is a strong tendency for successive ripening such that the earlier maturing seeds on a single plant ripen and are shed before the later ones have developed. This loss before harvesting is often referred to as 'shattering' or 'shedding'. Other sources of seed loss before harvesting include birds, small rodents and inclement weather.

There is a wide range of bird-repelling devices and methods used throughout the world to prevent loss. These can be classed and described as:

1. Auditory: automatic acetylene 'exploders', rattles and 'humming line' (this is a length of polypropylene tape stretched horizontally across the crop, just above the height of the plants). A traditional method for scaring birds from crops in the Yemen is the use of the traditional sling; even without a missile, the sling makes a very loud cracking noise. Figure 4.1 shows a worker using the traditional sling for this purpose.
2. Visual: scarecrows (traditional but very effective), an anchored aerial balloon which often has a hawk-like silhouette hanging from it, with reflective tape which is red on one side and silver on the reverse.
3. Obstruction: netting is successful for the protection of small-scale seed production areas but is impractical over large areas. Nets designed to trap birds are used in some areas, especially when dealing with introduced or escape species, although this practice is illegal in some countries.

Lodging

This term is used to describe the collapse of a crop before cutting or harvesting. Crops for seed production are especially vulnerable from the start of anthesis when the extra weight of the inflorescences or seed heads makes the mother plants top-heavy.

In addition to the susceptibility of specific crops such as cereals to lodge after seed-head development, several cultural and environmental factors can contribute to the incidence of lodging: these include wind, high soil nitrogen regimes during early plant development, heavy rain which can either weigh plants down or reduce efficiency of root anchorage, and straying or wild animals.

Once lodging has occurred the plants tend to deteriorate and, depending on their development stage, may not regain their

Fig. 4.1. Creating the 'cracking' noise of the sling to scare birds from crops in Yemen.

normal vertical posture. This can result in a poor microenvironment in the crop canopy, insufficient pollination for cross-pollinating species and in relatively wet seasons, or areas, the seed quality may deteriorate with subsequent reductions in seed yield, vigour and viability.

Stage of Harvesting

The ripening process is interrupted if the seeds are harvested too early and seed quality may be adversely affected. The environmental effects before harvesting on seed viability were reviewed by Austin (1972). Studies of seed maturation and harvesting in some of the forage grasses have been made by Peglar (1976), who demonstrated that the germination of seed harvested 2 weeks following anthesis was high in Italian and perennial ryegrass, meadow fescue, tall fescue and cocksfoot. However, with timothy cultivars high germination potentials were not attained until 3 weeks following the peak of anthesis. The author further reported that endosperm development stages were found to be reliable guides to assessment of correct harvest date and suggested that examination of endosperm consistencies could be used to corroborate assessment of harvest times.

Decisions on stage or time to harvest the *Trifolium* species (the clovers) is another example of a crop with a very extended flowering period (anthesis) and therefore ripening over a long period with the earlier inflorescences setting seed before the later flowers have even developed; the dilemma is further aggravated by the fact that the clovers generally shed their mature seeds very readily.

Generally, the later the harvest of many crops the higher the seed yield, but as illustrated in Fig. 4.2 losses increase as harvesting is delayed once the optimum percentage of seeds reach maturity. The onset of inclement weather may well have a deleterious effect on a seed crop which is nearing the optimum harvest stage. Thus for any individual crop the ideal harvest time is immediately before the loss of mature seeds exceeds the amount of seeds yet to reach maturity.

The incidence of shedding from ripe material is increased during dry weather. Crops which are particularly prone to shed their seeds while being cut are often best handled at times of comparatively high relative humidity. In arid areas this can be early in the day when the effects of overnight dew are still effective, after rain or even following irrigation in some crop species.

Accelerating the Drying of Crops Prior to Seed Harvest

In some seed crops it is helpful to reduce foliage and plant material when the seed is approaching maturity. This facilitates mechanical harvesting of the seed by reducing the amount of plant debris, increases the rate of natural seed drying, brings all seed material in the seed lot within a closer range of moisture content, avoids loss from windrows and will to some extent control weeds. There are three types of material that have been used for this purpose, desiccants, defoliants and soil sterilants, although desiccants are the most commonly and widely used. Hole and Hardwick (1977) have reported the results from trials under UK conditions for the drying of *Phaseolus vulgaris* plants prior to harvest using one or more of each type of material outlined below. Some of the chemicals are not available in all countries and the cost of material and its application has to be taken into account when evaluating the results, which also includes seed moisture content and likely seed damage.

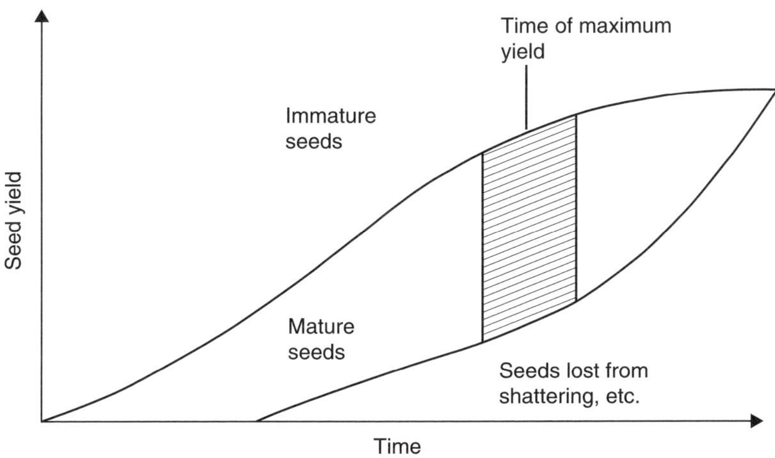

Fig. 4.2. Schematic representation of the interaction of seed maturity, potential yield and loss resulting from shattering and birds.

Desiccants

Examples of desiccants used in seed production are the bipyridyl herbicides, diquat and paraquat. The main seed crops to which they are applied are bird's foot trefoil, dwarf beans and *Beta* species. The use of desiccants is relatively expensive and there is a risk that they may cause some seed discoloration and/or a reduction in potential germination; they are unlikely to be acceptable for the production of organic seeds. However, their use can assist the rate and efficiency of subsequent direct combining and also reduce the moisture content of seeds by the time they are harvested.

Defoliants

Ethephon is an example of a defoliant used in seed production. Its mode of action is to trigger the release of ethylene, which causes leaf abscission.

Soil sterilants

The principle of the use of soil sterilants to increase the rate of maternal seed-bearing plant drying was that the plants roots are chemically damaged and water uptake is stopped. However, although the theory of this may be promising, the chemical residues may be detrimental to the following crops. Metham-sodium has been used for this purpose.

Harvesting

As illustrated in Fig. 4.2, the optimum time to harvest is a balance between immature seeds yet to ripen against loss of ripe seed, i.e. estimating the optimum time to achieve maximum yield of good quality mature seeds. In addition the method of harvesting and environmental conditions at the time of harvest can have very significant effects on the ultimate seed quality and yield.

Indications of seed crop maturity

The general signs of maturity in a seed crop include:

- Loss of green pigmentation in the mother plants.
- Change in seedcoat colour as ripening progresses.

There are some crop species whose progressive seed ripening is extremely slow, e.g. *Vicia faba* (field bean, broad bean, faba bean) in which the plant turns black with accompanying leaf loss prior to seed maturity; this contrasts with some grass species which still have some green pigmentation of their stems when the seeds are ripe. The optimum seed harvest stage is given, where appropriate, for individual crop species in Part 2 of this volume.

The most important objectives to be considered or achieved when planning or timing seed harvest should include:

- Purity of the seed lot harvested.
- Maximum maturity of the seed, assessed by the 'thumb nail' test (determined by pressing sample seeds with a thumb nail; the seed should be at least doughy but not so hard that it cannot be marked with the thumb nail).
- Appropriate moisture content for the crop species (determined in the field by rubbing out a seed sample and testing the moisture content).
- Maximum seed yield.

Harvesting systems

There are two basic systems for seed harvesting.

1. Cut the crop and allow time for further drying either *in situ* or under cover. The crop may be left spread out or left in windrows for further drying or removed for further drying off the field. The latter system is advantageous for those crop species that have a prolonged anthesis and seed development time. Other advantages of this system are:

(i) The seed crop can continue maturing on the mother plants;

(ii) Harvesting can start earlier, although it may take longer to completion.

2. The seed is separated from the maternal straw *in situ* and taken elsewhere for further processing; this may be referred to as combine harvesting.

Mechanized cutting

When the cutting operation is mechanized the material is either left *in situ* by a machine with a cutter bar or a machine that places the material in windrows. Machines capable of this operation have a draper belt in addition to the cutter bar and the cut material is carried under the machine and deposited in a swath. The swaths can either be turned into windrows or left *in situ* to dry according to the density of the material and the rate at which they will dry in the field.

Combine harvesting

This joint operation is done by combine harvesters that cut and thresh the material in one process; however, it is important that the standing material is uniformly ripe. Combine harvesters can be used to harvest seed from a large-scale cereal crop or a crop which has lodged. Combine harvesters can also be used as stationary threshers or for collecting and threshing material already cut and which has been left in windrows for further drying, in which case a pick-up head or reel is fitted.

Harvesting by hand

Hand harvesting is still done for very high value seeds, when the total area to be harvested is very small (e.g. breeders' and basic seed) or in areas where there is adequate hand labour. Seed heads, dried fruits or other forms of modified inflorescence containing the mature seeds (e.g. maize cobs) are picked or cut with knives or secateurs into baskets or other suitable containers. In some crops which are cut by hand a larger part of the plant is removed with the seed heads. This is achieved, for example for small areas of brassicas, by the use of knives, machetes, sickles (reaping hooks) or in the case of some crops, such as beans, by pulling up the whole plant. Hand-harvested material is usually either dried further on tarpaulins or other suitable sheets; the cut material is then placed in suitable structures on clean concrete floors which have a smooth surface or in airy racks or boxes.

Threshing

Dry seeds are removed from the mother plant material by flailing, beating or rolling.

It is important to ensure that unnecessary fragmentation of plant material does not produce debris which is either difficult or costly to separate from the seed sample during subsequent processing; it is also important to avoid mechanical damage to the seeds.

Hand threshing

Hand threshing is a relatively cheap method for small seed lots and is still used in some countries for large seed lots where hand labour is available. Several hand methods are possible including rubbing, beating the material against a wall or the ground, or flailing.

Seeds which have been hand threshed are usually still mixed with the plant debris and further separation is done by winnowing and/or sieving.

Machine threshing

The main feature of threshing machines is a revolving cylinder in a concave. The cylinder is driven by a motor or engine and is capable of reaching 1200–1500 revolutions per minute (rpm). The higher speed is suitable for agricultural grain but as low as 700 rpm is required for the large-seeded legumes. There are stationary and mobile threshers.

Both the cylinder and the concave have spiked steel teeth, angle bars, rasps or rubber bars; threshers with bars are more frequently used for vegetables than models with teeth. The concave opening is adjusted to allow the free passage of seeds, which are collected in a container, usually below the drum. Some threshers have interchangeable concaves for use with large-seeded legumes.

Some types of thresher incorporate sieving, screening or aspirating components to assist in the initial separation of seeds from plant debris. These modifications or additions can include the possibility of adjusting cylinder speed, cylinder clearance, concave mesh, airflow and screens. These refinements are essential for dealing with a range of different crop species.

The possibility of surface and sub-surface damage to seeds is increased if the cylinder speed is too fast, the cylinder clearance is too narrow or the mesh of the concaves is too small.

Seed Processing

The term 'seed processing' is used by the seed industry to include a wide range of operations to improve or *upgrade* seed lots after threshing. The processing is done so as to ensure that the seed which ultimately reaches the farmer is of the best possible quality. The objectives of processing may include removal of a wide range of materials, some of which will hold moisture to a detrimental effect during storage or distribution. The undesirable materials can include:

- plant debris including chaff and awns;
- non-plant material such as soil, sand, stones and rodent faeces;
- seeds of other crops and weeds;
- seed appendages which would otherwise interfere with the free running of the seeds;
- mechanically or insect damaged, discolored or immature seed; in some instances the seeds may be viable but with low germination potential and therefore of no commercial value, especially as they will adversely affect the germination test results of the overall seed lot; and
- seeds which are not within the size range or density of the seed lot.

The seed lot

A *seed lot* may be defined as a quantity of seed of one cultivar, of which the history, origin and weight are known. The specific seed lot can be:

- seed harvested from one crop;
- a blend of seed harvested from different crops (but the same cultivar) on the same farm; or
- a blend of lots from different farms, each constituent lot being the same cultivar; in this case a Certification Authority should give its approval.

The weight of a seed lot is normally restricted to 20t for cereals and pulses, or 10t for smaller seeds (i.e. smaller than wheat). These weight specifications for seed lots are the maximum normally covered by an international certificate.

Cleaning and upgrading the bulk seed lot

Generally there is a succession of operations in three stages, these are:

- pre-cleaning (also referred to as *conditioning* or *scalping*);
- basic cleaning; and
- separation and grading.

All seed lots do not necessarily have to go through all three stages; e.g. a seed lot may be sufficiently free of non-seed debris that it can proceed to basic cleaning without precleaning.

Winnowing

After threshing, the dried seeds can be separated from less dense or lighter debris by air movement, usually referred to as winnowing; the operation can be done by hand or by machine.

The simplest method of winnowing is to hold a tray of seed with both hands and toss the seed in the air; the low density materials, including chaff and light seeds, are carried away in the breeze while the denser material, mainly good seeds, fall back into the tray. This method is still used by subsistence farmers and for dealing with very small seed lots (see Fig. 4.3). The alternative small-scale method is the use of various sized sieves. Hand winnowing and/or sieving are the only methods available where there is no suitable power supply. The further development of this principle has led to the design of machines that combine aspiration and sieving.

Pre-cleaning ('scalping')

During pre-cleaning the bulk of plant debris and any other non-seed materials are separated by vibrating or rotating sieves. In some species seed clusters are also separated during pre-cleaning. The pre-cleaning machines usually have an airflow to remove materials lighter than the seed. Seed lots are usually pre-cleaned before drying.

Fig. 4.3. Traditional hand winnowing.

The seed processing unit

The unit for seed processing consists of a selection of machines, some of which are suitable for a wide range of crops, as with air/screen cleaners, while others are designed for individual crop genera, e.g. the maize sheller, or specific problems arising in quality control, e.g. the needle drum separator; each of these machines is described below, along with other specialized seed processors. The optimum layout and flow of materials for seed cleaning is shown diagrammatically in Fig. 4.4, although the sequence and types of machine in the line can be modified according to the range of crop species to be dealt with.

The separations of seeds from other materials are based on physical differences; these include size, shape, length, density, surface texture, colour, affinity to liquids and relative conductivity. It is generally agreed that processing costs time and money. The skill of the processor and their choice of machine for maximum purity of the finished seed lot is to get the best quality final seed lot with the minimum of machine operations.

Choice of machine depends on:

- the crop species;
- the stage of processing; and
- specific problems with an individual seed lot after initial cleaning.

Separation and upgrading

These are normally the final processes, which improve the mechanical purity of the seed and may also be done to remove a specific contaminant from the seed lot which was not achieved by the screen and air cleaning operation.

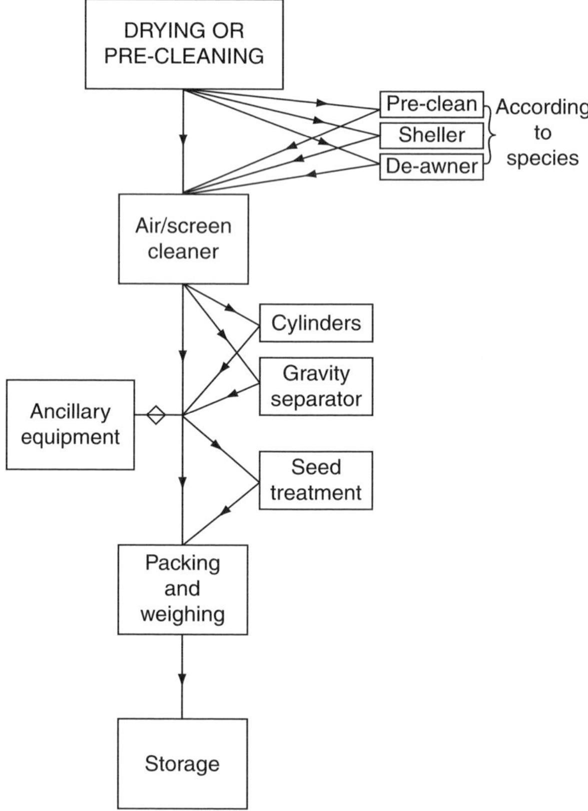

Fig. 4.4. Plan of a typical seed processing line.

Descriptions of seed processing machines designed for specific purposes

Maize sheller

Maize seed is usually removed from the dried cobs by subjecting them to a shelling action; seed damage is minimized if the shelling is done after the cobs have dried; the seed moisture content should not exceed 14% when shelling takes place. As the cobs pass through a drum which has a rotating beater with pegs, the seed is separated through a concave screen. Some machines incorporate fans to remove dust and debris from the cobs. The designs of individual machines will depend on potential throughput and whether or not it is for fixed or mobile installations (see Fig. 8.1). Care must always be taken to ensure that the type of machine used and its setting do not damage the seed.

There are hand-held maize shellers: these are more commonly designed for domestic use in food preparation, but may be the only shelling device available to many subsistence farmers especially where fuel for machines is not available or for very small harvested lots where genetic purity is important.

De-bearder

The de-bearder (also known as a de-awner) is so called because it is used extensively for de-bearding barley seed. It is also used for de-awning carrot and dill seed. This machine is used for these crops following initial pre-cleaning. This processing machine is essentially a drum with horizontal arms rotating inside. The arms rub the seeds against the drum and thereby remove appendages, or the arms can separate seeds which are in clusters.

Spiral separator or spiral gravity separator

The spiral separator is used for separating non-spherical or irregular-shaped seeds from a round-seeded species, for example to separate complete (undamaged) *Brassica* or legume seeds from split or broken seeds. This type of separator has a minimum of two vertical spirals around the vertical axis.

The seed material to be processed is put at the top of the inner spiral column; the more rounded seeds have a greater velocity and travel down the outer spiral while the flatter and irregular-shaped seeds travel down the inner spiral. The separated lots are collected at separate outlets known as 'spouts' at the bottom.

Disc and cylinder separators

Generally in the modern seed processing industry these types of separator have replaced the spiral separators. They operate on the principle that one fraction of a seed lot is picked up in small depressions in a disc or inner surface of a cylinder while the other fraction remains loose and is thereby separated. An example of their application is to separate pieces of stem from seeds and for grading some seed crops. The external view of an indent cylinder is illustrated in Fig. 4.5.

Gravity separator or gravity table

These machines separate according to an individual seed's density. They are capable of separating sound seeds from other materials, including seeds that are mechanically damaged, diseased, light-weight, sterile or insect damaged. The seeds are fed onto a vibrating and sloping deck above a plenum chamber; air passes through the deck which is covered with either a porous cloth or fine wire mesh. The deck's vibration, lateral and lengthways slopes are all adjustable, as also is the air distribution. There are two basic types of gravity separator, according to whether the deck is rectangular or triangular; Fig. 4.6 illustrates one with a triangular deck. The different grades of material gravitate in different directions on the deck and are each collected separately. The separation of the fractions into horizontal layers during the machine's operation is usually referred to as stratification.

Roll mill, also known as a dodder mill or velvet roll mill

This machine is widely used to separate rough-coated weed seeds such as dodder and cranesbill from smooth seed e.g. clover

Fig. 4.5. General view of a pair of indent cylinders.

or lucerne. With good operating skills the same principle can be adopted for separation of larger seeds such as removing wild oats from cereals. The principle is based on the surface textures of seedcoats, rough from smooth. The machine has two parallel and inclined velvet-covered rollers or cylinders. The covered rollers revolve outwards from each other. The seeds are fed in at the higher end of the pair of rollers. The smooth seeds gravitate down the trough formed by the two rollers and are collected at the bottom. The contaminating seeds and other materials with rough surfaces adhere to a roller's velvet surface and are deflected into a discharge spout. The roller's speed and feed rate are adjustable according to the specific requirements of the seed lot which is being processed.

Magnetic separator

This machine is used for separation and depends on the differing surface characters between the two fractions. The seed lot to be upgraded is first treated with iron dust, moistened with water or oil, which adheres to the rough coated fraction, e.g. seeds of the common weed cleavers (*Galium aparine*) contaminating smooth seed-coated material, e.g. *Brassica* seeds. The material is then passed over magnetized rollers and the clean seed is collected at the side while the magnetized material which has picked up iron dust is brushed off separately.

Electronic colour separator

These machines are used for the separation of 'off-colour' seeds, which may be diseased, from an otherwise clean seed lot. Examples of their use include removal of individual discoloured or stained pea or bean seeds or any which are infected with halo blight. The seeds are fed by belt, gravity or roller past a photo-electric cell which triggers a jet of air to remove the off-colour seed from the main seed lot. Both monochromatic and bichromatic instruments are available to enable different colours as well as different intensities of the same colour to be detected and removed.

Precision air classifier

This type of machine separates materials with a range of sizes and specific gravities by floating them through a rising air-stream. The machine works on the principle of differential specific gravities between fractions, which can be separated by fine adjustments to the machine.

Fig. 4.6. Gravity table with a triangular discharge deck.

Needle drum separator

Seed of peas and beans which have damage caused by weevils can be separated from sound seeds by a needle drum separator. Separation is made by a series of needles housed on the inside of a rotary drum. Seeds are transferred to the revolving drum, and those seeds which have weevil holes become attached to the needles and are separated from the sound seeds.

Seed Drying

The moisture content of a seed crop which is at or approaching maturity is used as an indication for the timing of harvest. Once the final processing of a seed lot has been completed it goes into storage, either in bulk or packaged, but before any form of storage takes place it is essential that the seed moisture content is determined and reduced if indicated. The moisture content is usually done with a seed moisture meter, although the most accurate method of determining the moisture seed lot is the method described by ISTA (2009a). There is often no need for further drying when seed has been produced in arid conditions unless some weather mishap has befallen it. There are safe storage moisture levels for individual crop species for open storage and also storage in vapour-proof containers.

Seed that is above its optimum moisture content has to be further dried to reduce it to the required moisture level. The optimum safe moisture level for storage varies between crop species. A useful guide is 14% or less for short term storage and 10% when storage is for in excess of approximately 6 weeks. There are several drying systems, which are described below.

Seed drying systems

Maize cribs

This is a relatively low-cost traditional drying method for drying maize cobs. A crib is usually constructed by inserting stout wooden posts vertically in the ground in a rectangular pattern, with a wire netting surround to form the crib or cage. The prevailing wind velocity and direction are taken into account. Longer, narrower cribs are used in areas with a higher relative humidity (RH) whereas in areas of low RH the cribs are near to square in plan; a rule of thumb is that the crib's height should not be greater than 1–1.5 its width. Figure 4.7 illustrates a maize crib in Kenya; shelled seeds are shown being open air dried on a sheet in the foreground.

Solar driers

Although not truly traditional, the evolution of the solar drier has developed from drying seeds or seed-bearing material in the open air following its harvest. The solar drier is an energy saving batch drier, see below for other types of batch driers. Earlier models were built *in situ* using a wooden frame covered with black polythene. More recent manufactured versions consist of an inverted pyramid-shaped chamber made of black mild steel. There is a vertical chimney from the apex which exhausts to the exterior. The chamber is heated by solar radiation (Figs 4.8a,b). The material to be dried is placed on an inside tray just above ground level. As the chamber warms up the interior, moist and warmer air rises and passes up the chimney. Access to the drying chamber is through a side or front closable opening, so that the tray and seed material can be reached and checked for progress. The dryer has a very useful role in suitable climates, especially for groups of farmers in remote locations with no power supply.

Batch driers

The batch driers operate on the basis of warm air being blown through a stationary seed layer until the seed moisture level has been reduced to the required level. This type of drier is very suitable for on-farm drying and also relatively small volumes of seed; the three most commonly used types are:

1. Horizontal drier: in this design of drier the seeds are in a confined space, such as a box structure with a slatted floor through which warm air is blown. On a larger

Fig. 4.7. Maize crib, with 'shelled' maize seeds in the foreground. Photograph by kind permission of The Real IPM Company (K) Ltd (www.realipm.com).

version of this system, air ducts can be laid on a floor of a barn or other suitable building or structure. The seed is piled over the ducts to be dried.

2. Sack drier: this is similar to the horizontal drier except that seeds are in woven sacks which are placed on the grid above the warm air ducts.

3. Vertical drier: this consists of two perforated concentric cylinders. The space between them is filled with seed; air is blown into the inner cylinder and passes through the mass of seed.

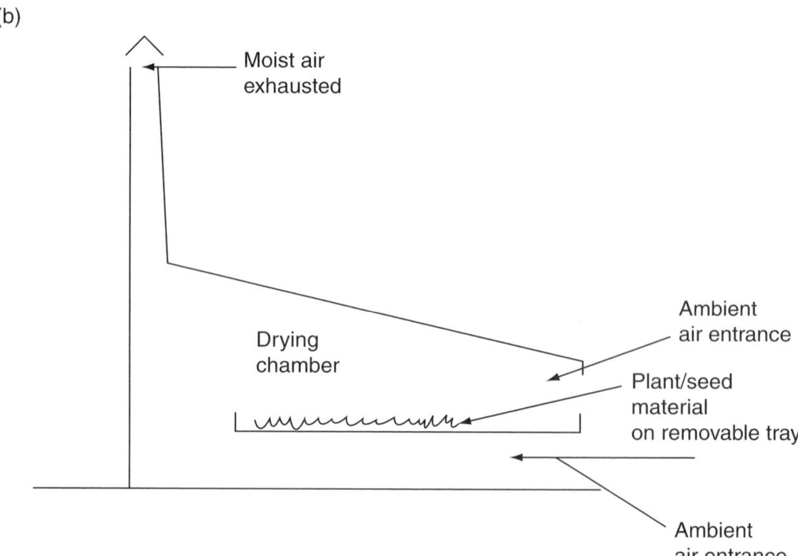

Fig. 4.8. (a) A solar drier in Fiji, constructed with a timber frame and black polythene cladding and (b) side elevation of a solar drier.

Further Reading

Coleman, J.D. and Spurr, E.B. (2001) Farmers' perceptions of bird damage and control in arable crops. *New Zealand Plant Protection* 54, 184–187.

Gregg, B.R. and Billups, G.L. (2009) *Seed Conditioning*, Vol. 1: *Management*, Parts A and B. Taylor and Francis, New York.

Gregg, B.R. and Billups, G.L. (2010a) *Seed Conditioning*, Vol. 2: *Technology*. Taylor and Francis, New York.

Gregg, B.R. and Billups, G.L. (2010b) *Seed Conditioning*, Vol. 3: *Crop Seed Conditioning*. Taylor and Francis, New York.

ISTA (1963) Drying and storage. *Proceedings International Seed Testing Association* 28, 4.

ISTA (1973) Seed storage and drying. *Seed Science and Technology* 1, 3.

ISTA (1987) *Handbook for Cleaning of Agricultural and Horticultural Seeds on Small-scale Machines*, Part 1. International Seed Testing Association, Zurich.

ISTA (1988) *Handbook for Cleaning of Agricultural and Horticultural Seeds on Small-scale Machines*, Part 2. International Seed Testing Association, Zurich.

ISTA (2009) *International Rules for Seed Testing*. International Seed Testing Association, Zurich.

5

Seed Storage

Definition of Storage Period

The storage period of a seed lot should be defined as the total time from seed maturity through to sowing. The duration may be relatively short, perhaps even only weeks, but it is also possible that seed lots may have to be stored for several years depending on the seed class or especially if it is breeders' stock seed. Although a seed lot can undergo several operations such as cleaning and packaging, or a waiting period while held in stock by a retailer or farmer before being sold or sown, it is important that this total time from maturity on the mother plant through to sowing be regarded as the overall storage period; it is throughout this entire time that seed is subject to any adverse environmental influences. The stages comprising the total possible storage duration of a seed lot include some, or all of the following:

1. Post maturity drying.
2. Seed extraction or threshing (outside or under cover).
3. Seed cleaning (outside or in buildings).
4. Holding in seed store or warehouse.
5. Packaging (usually in a purpose-built structure).
6. Transit and distribution (all modes of transport, handling stages transport distances).
7. Marketing (wholesale and retail centres).
8. Point of distribution or sale (seed store, bazaar or farm supplier).
9. On-farm storage (from receipt of seed at the farm or smallholding to sowing).

When seed extraction and drying have been completed it is necessary to store the seed under the best possible conditions to ensure that the maximum potential germination and other seed quality factors are maintained. Stored seeds are the primary input of a country's cropping programme and are vital links between successive crop generations. In commerce stored seed represents a significant proportion of the seed company's financial and genetic assets while for the subsistence and small farmers it is often the means of providing food and/or income for their families and dependants in the following growing season.

Reasons for Storage

There are many reasons for storing seed of individual crop species beyond the next possible sowing season. In some cases it may be uneconomic for seed organizations to multiply each seed stock annually; reasons for isolation in time have been discussed in Chapter 3. In addition, the recurrent annual cost of multiplying each cultivar offered by a seed company or seed marketing organization has to be considered. Seed yield is influenced by many factors, including the multiplication rate of the species and cultivar, provenance and seasonal variations during development and pre-harvest ripening. It is not always possible to make accurate forecasts of yields; thus satisfactory storage is also a useful way of ensuring that surplus seed is kept in satisfactory condition for future use.

Storage for fluctuating demand and seasonal variations

Farmers' and growers' demands for seed of specific crops can depend to some extent on the general market economy, market trends in crop demand or market successes or failures in previous seasons. In some countries, or groups of countries, the introduction or withdrawal of subsidies can alter farmers' preferences for growing certain crops. Satisfactory seed storage assists the seed industry to buffer these variations in requirements. In some instances specialist seed producers are in a different area of the world from the consumer and stocks may arrive at their destination out of season. The viability of seeds can be adversely affected during the period of transit and awaiting sowing if they are not stored properly. It is also important to ensure that stored seed is protected from hazards such as insects, vermin, fungi, fire and pilfering.

Storage of valuable stocks by seed organizations

All seed stocks have a value, which is related to the level of genetic purity and multiplication rate. Breeders' and basic seed stocks are normally stored under as ideal conditions as possible to reduce the frequency of their multiplication; this makes the maximum use of each seed stock and keeps the number of multiplications to a minimum.

Storage of germplasm

The storage of seed as germplasm, a potentially valuable genetic resource, calls for very long-term storage of relatively small seed lots. Germplasm centres, usually referred to as gene banks, use specialized storage methods which have been described by Cromarty *et al.* (1982). The operating standards for conservation of orthodox seeds in gene banks have been described in an FAO publication (Anon., 1994).

Natural longevity

The pre-storage history of a seed lot will also have a very important influence on its subsequent storage qualities. Those factors affecting storage potential include mother-plant environment during plant growth and seed development. Ellis and Roberts (1980) have discussed and described a method for the prediction of seed longevity.

Pre-harvest field factors influencing storage potential

Micro- and macronutrient deficiencies during plant growth and development of a seed crop can have a major effect on seed storage potential in that initial germination potential is low and, although such extreme conditions do not normally prevail in commercial seed production, the levels of major nutrients have been shown to indirectly affect seed storage.

During seed production many of the environmental effects in the field which predispose the seed to a relatively short storage life are especially severe in the tropics and subtropics, but they may also occur in temperate regions. It is probably those arid regions that have adequate control over irrigation frequency and quantity that have the best pre-harvest quality control in relation to potential storage.

Field deterioration of seed prior to harvest is frequently associated with environmental factors such as high relative humidity, excessive rainfall and high temperature; combinations of some or all of these factors result in a reduced potential storage life of seeds. Some of the deleterious effects can be successfully counteracted by operations such as mechanical harvesting and seed drying. The effects of environment before harvesting on viability were reviewed by Austin (1972). The importance of microflora and airborne inocula during seed production in the field and storage fungi with their possible deleterious effects on stored seed, especially the cereals, have been discussed by Christensen (1972).

Pre-storage mechanical factors

Mechanical damage to seed during operations such as harvesting, processing or drying can reduce the potential storage life because the damaged seeds lose vigour sooner than undamaged ones. In addition, damaged seeds are more vulnerable to storage pests and pathogens. There are various forms of mechanical damage to seed which can occur before storage, one of which is threshing injury, e.g. cotyledon cracking of *Phaseolus vulgaris* (dwarf or French bean), and seedcoat abrasion of many species often as a result of incorrect cylinder speeds. Damage can also occur during processing if the seed is immature; the other extreme is over-drying, which produces brittle seeds that are predisposed to cracking or breaking.

The importance of seed quality at start of storage

The different components of seed quality are discussed in Chapter 6. Those that have a direct influence on the storage potential of a given seed lot are viability and vigour. In this respect the storage potential refers to the stage of deterioration as expressed by seed vigour. The seed lots with the higher vigour will be more able to withstand the various stresses derived from the storage environment throughout the entire duration of storage.

Effects of storage environment on seed

The two most important environmental factors which can affect seed quality during storage are temperature and relative humidity. In practice the ambient relative humidity plays a major role, first because the seed's moisture level is a function of relative humidity, and second the incidence of storage fungi and insect pests is largely influenced by the relative humidity of the microclimate within the seed mass.

Moisture Content of Seed

The seed most suited for storage has a moisture content that is not greater than 10% of seed weight. A seed's metabolic rate is extremely low, often undetectable. When in this state it is hydrophytic and is capable of taking in water, even from the water vapour of the atmosphere. Thus, however much care has been taken to lower its moisture content by drying before storage, it will quickly take up water again when stored in ambient conditions.

Combined temperature and relative humidity

The loss of seed viability is slower at lower temperatures than at relatively high ones. In practice it is the combined effect of temperature and relative humidity that reduces potential viability or longevity of seed throughout its storage life. Many areas of the world have periods of fluctuating temperatures coupled with periods of high relative humidity, the combined effect of which influences the rate of seed deterioration during short periods of uncontrolled storage. Harrington (1963) provided two useful 'rules of thumb' for relative humidity and temperature, i.e.:

1. For each 1% reduction in moisture content the storage life of seed is doubled.
2. For each 5°C lowering of the storage temperature the storage life of the seed is doubled.

The relationship between storage temperature, seed moisture content and the viability period of the seed has been discussed by Roberts (1972). The construction of viability nomographs for use as a guide to the prediction of viability in hermatic storage and to a limited extent under open storage conditions have been described by Roberts and Roberts (1972).

The extent to which it is decided to control both temperature and relative humidity will depend on the value of the seed and the climate of the proposed storage location. In some arid areas of the world the seed's potential germination is not reduced significantly during short periods of uncontrolled storage. This is because the natural drying of seed is satisfactory following seed maturity in the field and there is a sufficient low relative humidity during the subsequent storage period.

The survival potential of seed at a range of temperatures and moisture contents has been quantified by Ellis and Roberts (1981).

Table 5.1. The typical moisture contents (%) of seeds in equilibrium with a relative humidity ranging from 30% to 70% (after Thomson, 1979).

	Relative humidity (%)				
Crop	30	40	50	60	70
Wheat and rye	9	10	12	13	15
Barley and oats	8	10	11	12	14
Maize	9	11	12	13	14
Sorghum	9	10	11	12	14
Rice	8	9	11	12	13
Ryegrass	9	10	11	12	14
Lucerne	8	9	10	12	14
Dwarf bean	7	10	12	14	16
Pea	8	10	12	13	15
Beet	7	8	10	11	13
Soybean	7	7	8	10	12
Cole crops	5	6	7	8	9
Groundnut	4	5	6	7	9

Subsidiary effects of relative humidity

When the seed moisture is between 40% and 60% germination will occur, resulting in the death of the seed's embryo when it occurs during storage. In addition to the direct effect of the seed's storage life, high moisture content has indirect effects, especially on the microflora and microfauna of the seed store environment. Most storage insect pests' activities, including reproduction, are stimulated with seed moisture above about 8%. Furthermore, the growth of fungi on seed will commence when the seed moisture content exceeds 12%. At seed moisture content above 18–20 %, local heating can occur as a result of total metabolic heat of the biomass; this in turn can be responsible for reducing viability or even spontaneous combustion. The typical moisture contents of seeds in equilibrium with a relative humidity range from 30% to 70% are given in Table 5.1.

Storage Environment

The storage environment can be influenced in two main ways:

1. The building (or structure) in which the seed is stored.
2. The temperature and relative humidity.

It is important that the seed is prepared for storage in a purpose-built structure as quickly as possible after harvesting; and that a system is organized so that the seed remains in the storage environment for as long as is practical prior to distribution for sowing.

Seed Store Construction

Seed stores should be designed to maximize security, to minimize fire risk, to exclude birds and rodents, and to keep the entry of insects and microorganisms to a minimum.

The ideal building should have a raised and smooth-finished reinforced concrete ground floor with a rodent-proof lip. Entry can be by removable ramp and the raised floor designed to match up with the loading level or height of delivery vehicles. The roof should be pitched and overhung to offer the best possible runoff of storm water and to provide shade and extra protection for the ventilator openings. There should also be a pitched canopy roof over the entrance chamber.

The extent to which rodent proofing, measures to deal with excessive storm water (including surface runoff water from higher surrounding areas) and extremely high temperatures are included as design features will depend on the location, topography and prevailing conditions throughout the year.

A double entrance door system incorporating an entrance lobby or antechamber should be included in the design, and ideally there should be no other openings or windows except those connected to environmental control systems.

The walls and ceiling should be constructed of smooth finished stone, mortar and concrete, which are lined with a moisture barrier of tar, aluminium foil or polythene. Wood should either not be used or kept to a minimum, as in time it can be attacked by rodents and other pests and present maintenance problems. The final finish of the interior walls should be of a material which will protect the insulation from damage made by trolleys, pallets or other handling systems, transport or storage devices. Any ventilation openings which are incorporated must be efficiently screened to exclude insects. The overall interior finish must be smooth with electrical conduit channels and other cracks smoothly sealed.

External finishes, especially the roof, should aim to minimize absorption of solar heat and to exclude water. Particular attention to these specifications and details are necessary in parts of the world subject to high temperatures and/or high rainfall. Seed stores should be provided with an adequate lighting system but with a low heat emission. There should be a regular cleaning and maintenance programme to ensure that all the design features of the structure are upheld during the life of the store.

Additional features of conditioned seed stores

The design and materials used in the seed store's construction should minimize the absorption of solar radiation and act as an effective vapour barrier. In some parts of the world these are the only features necessary to ensure a satisfactory storage life of seed. However, in many other areas further control of the storage environment must be achieved by air conditioning. This becomes a major requirement where the climate is such that design of the seed store is insufficient to ensure satisfactory potential life for the seed.

The actual design and structure of conditioned stores follows the specifications outlined above. But where conditioning is required it is extremely important to ensure that the best use is made of insulating materials in order to achieve the maximum efficiency from temperature-controlling equipment and value from operating costs. This factor has become very important because of the progressive rises in energy costs and the need for conservation of fossil fuels and protection of the environment.

Vapour proofing of the structure is also important and is achieved by building a continuous polythene film sealed with bitumen or other suitable sealant into all floor, wall and ceiling areas. This vapour barrier forms a seal completely enveloping the store and is installed on the 'wet' or exterior side of the heat insulation barrier. All doors must also be fitted with gaskets and there should be an air lock at the entrance.

An adequate power supply for operating the apparatus that is to control the store's environment is necessary. Control appliances should be positioned so that the heat they emit is exhausted to the store's exterior. An important consideration is positioning of air exhausts; hot air should be emitted from the store just below the roof line, while moist air should be expelled from near ground level if conditioned separately.

Temperature control

Storage temperature can be reduced by ventilation and refrigeration in addition to insulation and structural features. Complete temperature control is expensive in any part of the world and is unlikely to be used in commerce for seeds, regardless of their value. Total temperature control by refrigeration is, however, used in the long-term storage of germplasm collections and breeders' material and was outlined above, under 'Storage of Germplasm'. Work by Ellis *et al.* (1996) has indicated that storage at −20°C rather than 20°C is beneficial to seed survival, and that hermetic storage at 20°C of seeds first dried at 20°C to moisture contents in equilibrium with about 10% RH provides greater longevity than 5.5–6.8% moisture content in the five species studied (carrot, groundnut, lettuce, oilseed rape and onion).

Temperature control during the storage of 'short-term' and 'carry over' seed stocks is usually achieved by ventilation in conjunction with refrigeration, the degree of which will depend on the outside temperature.

Ventilation

The ventilation of seed stores should be considered in conjunction with the relative humidity of the ambient air, because it would be more harmful to the seed to lower its storage temperature if the result is to increase the seed's moisture content. Conversely, ventilation can be used to lower the storage temperature and the seed's moisture content when the relative humidity of the outside air is sufficiently low.

Storage engineers can design systems controlled either manually or automatically which operate ventilating fans according to the temperature and relative humidity of the outside air. Ventilation also enables a gentle airflow to be passed through bulk lots of stored seed as and when required, thus ensuring that hot spots do not develop that would endanger the stored material.

Another form of localized heating sometimes coupled with moisture migration can occur as a direct result of convection currents. These are most likely to occur in unattended stores with relatively poor insulation and are caused by drier warm air moving from a warm spot to a cooler one within the store. On cooling, moisture is condensed, which is subsequently absorbed by the drier seeds.

Refrigeration

The use of refrigeration in controlling seed store temperatures is generally confined to long-term storage of high value material, for example germplasm collections and breeders' stocks (Cromarty *et al.*, 1982). However, refrigeration is also useful in the tropics for valuable categories of other seed stocks.

Extra care and attention must be given to thermal insulation and structure of the store when refrigeration is to be included in the control systems.

There are four sources of heat within a seed store which the refrigeration has to cope with; these are derived from: (i) leakage from outside (despite insulation); (ii) field heat of seed and associated materials; (iii) respiration of seed within the store; and (iv) incidental heat derived from lighting, other equipment, workers and external heat that enters when doors are opened.

The cooling coils of a refrigeration unit should be situated within the storage area but the compressor must be sited so that its heat is given off to the exterior of the storage room. The relative humidity of the air is reduced during the refrigeration process. The moisture condenses on the cooling coils, which have to be defrosted at intervals. Although this reduction in relative humidity is an advantage, in practice the store's relative humidity is inversely proportional to its temperature and at temperatures below about 13.0°C the relative humidity is too high for open storage.

Dehumidification

An alternative system to refrigeration of seed stores is the use of a suitable chemical desiccant in a dehumidifier.

There are two types of chemical dehumidifier generally used in seed stores, the bed and the revolving drum. In each system the apparatus will add to the interior heat load if not carefully sited and it is therefore important that dehumidifiers be placed in the structure so that their heat is expelled to the outside of the store. Silica gel, which can absorb up to 40% of its own dry weight in water, is usually used for seed-store systems. In the bed system the silica gel is dried up to about 175°C to drive off all the absorbed moisture. After cooling, air from the storage area is blown through the dried silica gel bed. When the silica gel is again in moisture equilibrium with the air it is reheated to dry it before further re-use. Some bed systems use two beds per unit, in which case one bed is dehumidifying the store's atmosphere while the other is being dried to re-activate it. The operation of alternating beds is normally controlled by a time clock.

The revolving drum system has a desiccant bed that is divided between two air streams. Different sections are dried or used for absorption as the bed rotates. The revolving drum systems are capable of removing more moisture from a given airflow than the bed systems.

The choice of system will depend very much on local conditions and storage requirements, so in all cases qualified and experienced environmental control engineers should be consulted.

Storage in Vapour-proof Containers

There have been major developments that have led to the storage of seed in sealed moisture-proof containers. Most of the original research and development work with this technique has been done with seed of vegetable species because of the relatively small seed lots required by individual farmers and, with the exception of the larger seeded vegetable species in *Leguminosae*, the proportionately high value of vegetable seed per unit volume compared with cereals and most other crop groups.

The principle is that seed lots are dried to a moisture level slightly lower than they would be prior to normal open storage, and are then sealed in metal cans, packets or other suitable vapour-proof containers. As a result of this containerization or packaging, each seed lot is in its own sealed environment and may be stored at ambient temperature and relative humidity for 1 or 2 years, or even longer, with little or no deleterious effect on germination. In fact, provided that careful attention has been given to important details such as drying, and the determination of moisture content, and appreciating that different seed lots can differ in their composition, it is possible that the potential storage life of a seed lot at 8% moisture content can be doubled if 1% more moisture is removed before sealing.

Seed moisture before vapour-proof packaging

It is of fundamental importance that seed moisture content is reduced to a satisfactory and safe level before the seed is sealed in vapour- or moisture-proof containers. This is generally 2–3% lower than for other forms of storage or packaging in non-moisture-proof containers. The reason for this is that the atmosphere within a moisture or vapour-proof container will equilibrate to the moisture level that is in the contained seeds. This results in a long-term relative humidity that is too high for safe storage. For example, if maize seed with 13% moisture content is sealed in a moisture-proof container the enclosed storage atmosphere will equilibrate at a relative humidity of approximately 65%, which is too high for safe seed storage. In addition, some storage pathogens will become active within the sealed container and the seeds' respiration rate will be relatively high.

The conditions for safe storage of individual crop species or groups of species are given in Part 2 of this volume.

Seed-store Management

The level of hygiene within the store will have a long-term effect on seed quality and longevity. Only seed which has been through the final stages of processing should be taken into the store. All other materials should be excluded. In practice it is sometimes tempting to use seed stores for short-term retention of other plant materials such as small quantities of plant material awaiting seed extraction, but this misuse of the seed store can lead to the introduction of storage pests and pathogens. Plant materials other than seeds are very likely to add to the moisture content of the storage atmosphere. Other sources of water and moisture should also be excluded in order to discourage rats and other rodents.

A comprehensive programme for rodent prevention should be organized from the outset, rather than waiting for any control measures to become necessary later. The possibility of rodent infestation will depend on the method of containerization within the store as well as on the location and local conditions.

Rodent prevention and control programmes include the use of rodenticides such as the blood anticoagulants or other allowable proprietary poisons. The material should be used in accordance with the manufacturer's instructions and current national legislation relating to use of poisons and safety, as well as the safe disposal of dead rodents. Other methods to deter rodents from stores and warehouses include the installation of sonic sound systems which are undetectable by workers. The repetitions of sound recordings of birds of prey are also used as deterrents.

Systems for the reduction of potential germination rate of seeds kept in open stores have been described by O'Dowd and Dobie (1983).

Seed stores should not be used for storage or shelter of machinery, apparatus or any other materials not directly involved with the stored seed. Additional apparatus and materials make it difficult or even impossible to maintain a high degree of cleanliness. All surfaces should be kept clean and floors should be cleaned with vacuum cleaners in preference to brooms in order to minimize the build-up of dust. All waste materials should be removed from the store as soon as they are accumulated and disposed of by safe incineration, as far away from the store as is practical. Figure 5.1 shows the interior organization of a seed store.

Seed stores have a relatively high fire risk due to the dry nature of seed and the possibility of dust in the atmosphere. It is therefore vital that adequate fire prevention measures are formulated and every member of staff made familiar with them.

A system for entering, locating and retrieving seed lots should be adopted. The system should take into account the need for sufficient space between seed lots for access and air circulation. Small seed lots should be on suitable shelving and large quantities in bags or sacks should be neatly stacked on pallets.

The most sophisticated seed stores belonging to the larger seed companies have a computerized and fully automated retrieval system. But the main criterion for

Fig. 5.1. The interior of a seed store showing sealed bags and smaller seed lots with efficient use of pallets and crates.

any system, regardless of scale of operation or level of sophistication, is that all bags, cans or any other form of container must have a label inside and be clearly identified on the outside. Labels must be firmly attached; adhesive labels on tins or other containers should not peel off when subjected to the storage environment. The labels should be written in accordance with the inventory system adopted; this should maintain a record of each seed lot's year and other details of provenance, designation, samples tested, quantities removed and balance remaining. This information will then always tally with stock books and other records.

The structure and fabric of the buildings should be regularly inspected and any deterioration or damage must be restored immediately by competent tradesmen using appropriate materials.

Further Reading

Copeland, L. and McDonald, M. (2001) *Principles of Seed Science and Technology*, 4th edn. Chapman and Hall, London.

Cromarty, A.S., Ellis, R.H. and Roberts, E.H. (1982) *The Design of Seed Storage Facilities for Genetic Conservation*. International Board for Plant Genetic Resources, Rome.

FAO (2004) *Towards Effective and Sustainable Seed Relief Activities*. FAO Plant Production and Protection Paper 181, FAO, Rome.

Justice, J.O. and Bass, L.N. (1978) *Principles and Practices of Seed Storage*. Agriculture Handbook Number 505, USDA, Washington, DC.

Harrington, J.F. (1963) Practical advice and instructions on seed storage. *Proceedings of the International Seed Testing Association* 28, 989–994.

Roberts, E.H. (1972) *Storage Environment and the Control of Viability*. In: Roberts, E.H. (ed.) *Viability of Seeds*. Chapman and Hall, London, pp.14–58.

6

Seed Quality Control, Handling and Distribution

The quality of agricultural seeds is of paramount importance to the farmers and growers who sow them for the production of their next commercial crops or, in the case of subsistence farmers, to achieve food security for their dependants. Seed quality is a multifaceted concept, which includes a wide range of attributes.

Each of the following factors during the production and postharvest activities contributes to the overall quality of a seed lot:

- mother plant environment, including choice and suitability of site for multiplication;
- previous cropping history of the site;
- favourable and/or unfavourable field events and conditions from sowing throughout field production, ripening in the field and harvesting;
- crop monitoring and effectiveness of roguing and field inspections; and
- postharvesting drying, threshing, processing, storage and rate of deterioration during storage.

The quality of a seed lot is determined by specific procedures, which includes seed testing in specially equipped laboratories and growing on tests. These tests evaluate quality attributes such as germination potential, purity, health and moisture content. The vast majority of countries have adopted the rules for seed testing as specified by the International Seed Testing Association or those of the Association of Official Seed Analysts.

The International Seed Testing Association (ISTA)

The main objective of ISTA is the development, adoption and publishing of agreed and clearly defined standards for sampling, testing and promotion of uniformity and repeatability of seed testing procedures. To achieve this, the association produces the International Rules for Seed Testing that embrace the advice of its various technical committees. The Rules prescribe principles and definitions while the 'Annexes' describe methodology; these are published as supplements to the Association's journal *Seed Science and Technology*, to which the reader is referred in the Further Reading list and References, as appropriate.

The Association's secondary purpose includes the promotion of research in the subject areas of seed science and technology, and also the encouragement of cultivar certification. These objectives are achieved by organizing and participating in conferences and training courses.

ISTA has developed a system for the quality assurance of seed testing through its system of accredited laboratories. This embraces international standards and audit visits to ISTA approved laboratories, which ensure that requirements such as staff training, operating procedures, test records and laboratory independence are being adhered to.

In 2009 there were more than 70 member countries and over 1500 subscribers. The association produces an information pamphlet, publishes the *ISTA News Bulletin* three times a year and organizes a triennial World Seed Conference.

The Association of Official Seed Analysts (AOSA)

In North America the Association of Official Seed Analysts plays a similar role to ISTA. The association's constitution and bylaws were published in 1993 (AOSA, 1993). The history, legal framework, technical achievements and possible future developments of AOSA have been reviewed by Bradnock (1998a,b). AOSA publishes a journal, *Seed Technology and Seed Technology News*. AOSA and ISTA work in close collaboration on technical matters.

Seed Legislation

The general objectives of legislation related to seed are to ensure that seed which is offered for sale to farmers conforms to minimum required standards of quality, i.e. cultivar purity, health, vigour, germination and freedom from adulterating materials including weed or other crop seeds; in this context, moisture content is important as it affects the weight and genuine value of a given seed lot at point of sale. Coincidental to these objectives is the protection of farmers and growers from unscrupulous vendors. In countries where the seed industry is less developed the latter point may be more important, but once a seed industry is established, it is the monitoring of the different facets of seed quality that increase in importance, especially when seed certification schemes are introduced. This subject is discussed further by Kelly (1994) and national and international examples are detailed in Kelly and George (1998b).

Genetic quality

This may be referred to as 'trueness to type', cultivar purity or varietal purity. The strict control of seed generations or seed categories coupled with implementation and monitoring of seed certification procedures (or other monitoring schemes where they are applied) provide ways of assisting the authenticity of a seed lot.

Physical purity

The range of components of a seed lot can include seed of the stated species (i.e. pure seed), seed of other species (i.e. including other crop species and weeds) and contaminants. The contaminants may be derived from a range of sources including materials from seeds, parts of plants, living organisms (but not of plant origin) and other materials such as soil. The pure seed definitions of a wide range of crop species have been published by ISTA (2009b).

Seed health

This may be described as the extent to which seed-borne pathogens and/or pests are present. ISTA has published a list of seed-borne diseases with annotations regarding bibliographies and treatments (Richardson, 1990). The important seed-transmitted pests and pathogens are listed under individual crop genera or species in Part 2 of this volume. It should be noted that many of the species which can be seed transmitted can also be spread by other vectors.

Viability

This can be referred to as the potential germination and subsequent production of a seedling of the stated cultivar. Although the viability of an individual seed can be determined it is more usual to refer to the germination potential of a seed lot. In this context a seed lot is taken to be from a specific population; i.e. it is composed of a homogeneous population derived from the same stock, from the same production location at the same time and therefore having the same reference number; it is also assumed that all seeds in the seed lot have been treated in the same way including processing and storage. It is essential that methods used for determining the germination of a sample taken from a seed lot are both repeatable and reproducible, this is especially important from the aspects of legislation and international trade. The seed testing procedures described by ISTA (2009a) provide an international standard. The seed testing rules set by AOSA (2008) are used for example in Canada and the USA, although there is very close cooperation between AOSA and ISTA.

Vigour

Defined by ISTA (1995a) as 'the sum total of those properties of the seed which determine the level of activity and performance of the seed or seed lot during germination and seedling emergence'. The tests which are recognized are described or listed by ISTA (1995b) and AOSA (1983).

Moisture content

This is the percentage moisture of the seed lot. Although there are quick electronic systems in use, moisture content is usually determined by an oven-drying method as specified by ISTA (2009a).

Monitoring seed quality

The reasons for assessing the quality of a seed lot are numerous and include determining the monetary value, purity and moisture content. The different aspects of quality as defined in the section above are used to evaluate the sowing value of a specific seed lot, i.e. its true value to farmers and growers. The criteria may also be used to determine the commercial value, i.e. the price paid to the contractor who produced the seeds or vendor in a subsequent commercial transaction, and ultimately to protect the final purchaser of the seed who uses it to produce a crop. Different evaluations may be done in the pathway of the seeds from original producer through to the grower who sows them but each test, or evaluation, generally follows an internationally agreed and accepted procedure; this is essential for evaluations subject to legislation.

The criteria for the assessment of seed quality are referred to as *attributes* while the factors, or causes, which may predispose the seed lot to any of the attributes are known as *determinants*. It should be noted that there can be several determinants of both pre- and postharvest origin which may affect an attribute (Table 6.1).

The seed quality attributes are:

- germination and vigour;
- genetic purity (trueness to type);
- mechanical purity (physical purity);
- seed health; and
- moisture content.

Table 6.1. Summary of the effects of attributes relating to agronomy, handling, harvesting, processing and storage on the determinants seed quality.

Seed quality attributes	Pre- and postharvest determinants and/or operations influencing seed quality
Germination and vigour	Seed ripening conditions on the mother plant Postharvest ripening Drying Processing Avoiding mechanical damage Storage, including package and container environment Adverse and/or untimely seed treatments

Continued

Table 6.1. Continued

Seed quality attributes	Pre- and postharvest determinants and/or operations influencing seed quality
Genetic purity (trueness to type)	Accurate labelling at all stages and times Purity of original stock and/or basic seed Control of ground keepers (volunteer crops) Isolation (in time and distance) Roguing efficiency, at appropriate stages Application of cultivar descriptions
Mechanical purity	Processed seed free from seeds of other crop species, weeds or parasitic plant species Efficient winnowing, seed cleaning and correct use of appropriate specialist seed cleaning machines, depending on contaminants present in the seed
Seed health	Stock seed free of seed-borne pests and pathogens Adequate control of volunteer plants in the field which are alternative hosts to pests and pathogens Roguing out infected mother plants as soon as they are identified Timely and efficient seed treatments
Moisture content	Harvesting conditions Timely harvesting; drying and storage environment

Further Reading

Bruins, M. (2009) The evolution and contribution of plant breeding to global agriculture. *Proceedings of the World Seed Conference, Rome, 8–10 September 2009*. International Seed Federation (ISF), 1260 Nyon, Switzerland.

Douglas, J.E. (1980) *Successful Seed Programmes*. Westerview Press, Boulder, Colorado.

Gregg, B.R. (1983) Seed marketing in the tropics. *Seed Science and Technology* 11, 129–148.

Gregg, B.R., Delouche, J.C. and Bunch, H.D. (1980) Inter-relationship of the essential activities of a stable, efficient seed industry. *Seed Science and Technology* 8, 207–227.

Haldemann, C. (2008) ISTA and biotech/GM crops. In: *Seed Testing International*. ISTA News Bulletin No.136 October 2008, pp.3–5.

Hampton, J. (2009) Maintaining capacity in seed technology and seed testing. *Proceedings of the World Seed Conference, Rome, 8–10 September 2009*. International Seed Federation (ISF), 1260 Nyon, Switzerland, pp.187–195.

Kelly, A.F. (1994) *Seed Planning and Policy for Agricultural Production*. John Wiley and Sons, Chichester, UK.

Larinde, M. (2009) Building capacity in seed quality assurance in developing countries. *Proceedings of the World Seed Conference, Rome, 8–10 September 2009*. International Seed Federation (ISF), 1260 Nyon, Switzerland, pp.168–176.

Muschick, M. (2009) The evolution of seed testing. *Proceedings of the World Seed Conference, Rome, 8–10 September 2009*. International Seed Federation (ISF), 1260 Nyon, Switzerland pp.159–167.

Powell, A. (2009) What is seed quality and how to measure it? *Proceedings of the World Seed Conference, Rome, 8–10 September 2009*. International Seed Federation (ISF), 1260 Nyon, Switzerland, pp.142–149.

Van Der Burg, J. (2009) Raising seed quality: what is in the pipeline? *Proceedings of the World Seed Conference, Rome, 8–10 September 2009*. International Seed Federation (ISF), 1260 Nyon, Switzerland, pp.177–186.

Zecchinella, R. (2009) The influence of seed quality on crop productivity. *Proceedings of the World Seed Conference, Rome, 8–10 September 2009*. International Seed Federation (ISF), 1260 Nyon, Switzerland, pp.150–158.

7

Gramineae – Self-fertilized Cereals

There are five important genera in this group of self-pollinated cereals:

- *Triticum* spp. – wheat;
- × *Triticosecale* – triticale;
- *Oryza sativa* L. – rice, paddy;
- *Hordeum vulgare* L. *sensu lato* – barley; and
- *Avena* spp. – oat.

They are each discussed below along with their related species.

Wheat

The three *Triticum* species which are collectively referred to as 'the wheats' are:

- *Triticum aestivum* L. emend Fiori et Paol. – wheat, bread wheat;
- *Triticum durum* Dessf. – durum wheat; and
- *Triticum spelta* L. – spelt.

Origins and types

The wheats are thought to have developed from the cultivated einkorn which probably originated from wild subspecies of *Triticum* distributed in Iran, northern Syria and southern Turkey, northern Turkey and western Anatolia. The history and development of the wheats, *Triticum* spp., are described and discussed by Feldman (1984) in interesting detail. See also Further Reading for a more recent account of the origin and developments of wheat. The three *Triticum* species are the most widely cultivated small-grain cereals in the world that are cultivated for food production:

- *Triticum aestivum*: this is the foremost species used for bread making globally. It is hexaploid and is of significant importance in its traditional production areas, which have a continental climate. These areas include North America, India, north-central China, the Indian subcontinent, southwest Australia and Argentina.
- *Triticum durum*: the species used for pasta and also some bread production.
- *Triticum spelta*: cultivated on a relatively small scale; it is mainly produced as an animal feed, and also as a cereal in some mountainous areas of Europe. Although this crop is no longer widely grown it is regarded as having been a staple cereal crop in Europe from the early Bronze Age to medieval times. The seed remains enclosed in the glumes after threshing.

Each of the above *Triticum* species is also used for livestock feed.

Although the wheats are generally regarded as temperate crops they are produced successfully at higher altitudes in the tropics, largely as a result of plant breeding and the sowing of adapted cultivars. There is a very wide range of cultivars, many of which have been bred for specific environments and also with disease resistance.

For the classification of cultivars see UPOV Test Guideline 003 (see Appendix).

Triticum aestivum (Wheat, Bread Wheat)

Site

There is a high risk of wheat ground keepers arising from seed left from earlier crops, therefore there should be a minimum 2 year break from previous wheat crops unless the same cultivar and seed class (or a lower class) is being produced in which case the break time can be reduced. There should also be a break if other small seeded cereals, especially barley, have been grown on the same site; it is notoriously difficult to remove them during wheat seed processing. Wheat usually follows root crops, grain legumes or vegetables in the tropics and subtropics. The more ideal break crops should enable the field to be cleaned of species such as *Avena fatua*, *A. sterilis and A. ludoviciana* (wild oat species), and also serious weed species such as *Alepocurus mysuroides* (black grass). Most national seed certification authorities state the minimum standards to be met regarding weed seed and other crop seed tolerances in the seed lot.

Wheat succeeds best on well-drained fertile soils with a medium to heavy texture. Care must be taken to ensure that excessive nitrogen is not added during the site preparation or any subsequent top dressings as lodging will occur. The soil pH range for wheat is 6.0–7.5.

Pollination and isolation

Wheat is self-fertilized, therefore an isolation distance of 5 m is all that is normally required.

Seed crop establishment

Seed rates of 30 kg/ha can be used in very fertile conditions for the early stages of a multiplication programme; the resulting lower plant density increases the multiplication rate. It is very helpful for those workers who rogue or inspect the crop to use a wider spacing ('tramlines' or gaps) at intervals when sowing. Higher seed rates from 80 to 180 kg/ha are used in some regions for production of the lower seed categories. Drilling the seed is preferable to broadcasting.

Care must be taken if either straw shortening agents or herbicides have been applied as they may modify the plants' morphology and jeopardize roguing efficiency.

Roguing

There are two crop stages at which crops intended for seed production are rogued:

1. At ear emergence, when the seed has just started to fill.
2. When the glume and seed colour can be observed.

Harvesting

The seed crop is usually harvested by combine or 'stripper header' (this removes the seed heads, is faster and reduces the amount of separation required in the combine). The stage for harvesting has been reached when the straw has lost its green pigmentation and the seed is difficult to mark with a thumb nail. Seed can be damaged by combining when its moisture content is higher than 20%; similarly, seed with a moisture content below 8% can be damaged. The seedcoat is easily damaged outside these limits, resulting in reduced storage ability and germination. In small fields, or where there is plentiful hand labour, the crop can either be cut by hand and tied in bundles or cut by binders; the bundles are stood in stooks or stacks to dry further.

Unless the seed has been directly extracted during harvesting, as with a combine, the seed from the cut and dried bundles is either extracted by stationary combine or threshed by a more labour-intensive system; in this latter case the harvested seed is likely to require winnowing.

Drying

The seed is dried to 14% moisture content for short-term storage. Winter wheat, which is grown in the higher latitudes, has a very quick turnround from summer harvest to autumn sowing. The moisture content for this short period for winter wheat is only 16%. For all wheat types the long-term open storage moisture content is 10%, or 8% for vapour-proof storage.

The air temperature should not exceed 44°C when the initial moisture content is high, but increased to 49°C as the seed lot's moisture content reduces. Once the drying process has been completed the ambient air's relative humidity should not exceed 75%.

Seed processing

The seed can be further processed with an air/screen cleaner. Further processing can be achieved if required by a gravity separator and also by indent cylinders.

Triticum durum (Durum Wheat)

The above description for seed production of *T. aestivum* also applies to *T. durum*. However, the latter species tends to germinate prematurely and sprout in the ear, especially in high relative humidity; therefore, when prevailing conditions indicate that this is likely, the seed crop should be harvested before its moisture content falls below 18%.

Triticum spelta (Spelt)

This wheat is produced as described above for *Triticum aestivum*, a major character is that the seed remains enclosed in its glumes after it has been threshed.

Pests and pathogens

The main seed-borne pests and pathogens of wheat are listed in Table 7.1.

× *Triticosecale* Wittmack. (Triticale)

Origins and types

Triticale is a relatively new crop species derived from a cross between wheat and rye. The history of its development has been described and discussed by Larter (1984). There can be very significant differences in response to the environment by individual cultivars.

For the classification of cultivars see UPOV Test Guideline 121 (see Appendix).

Site

There should be a minimum break of 2 years after any small-seeded cereals, especially triticale, wheat, rye and barley; exceptions are usually allowed by national seed organizations when the preceding crop was triticale of the same cultivar produced from certified or otherwise authenticated seed.

Generally, the important weeds which should be controlled are as listed for preparations of wheat or rye seed production.

Pollination and isolation

Triticale is regarded as being liable to outcrossing because rye, one of its parents, is an 'out crosser'; however, the possibility and degree of cross-pollination differs between cultivars. The cultivars which are cross-pollinated should be isolated from other triticale and rye crops by 300 m for the production of Basic Seed or 250 m for Certified Seed production. The cultivars which are self-pollinated should be isolated from other triticale and rye crops by 50 m for the production of Basic Seed or 20 m for Certified Seed production.

Seed crop establishment

This is as described for wheat. Wild and escape grasses growing on the headlands and adjacent hedgerows should be kept

Table 7.1. The main seed-borne pests and pathogens of wheat; these may also be transmitted by other vectors.

Pest or pathogen	Common name
Alternaria triticana	Leaf blight
Alternaria spp.	Leaf spot, Alternaria blotch
Cladosporium spp.	Black kernel smudge, sooty mould disease
Gibberella zeae	Scab, head blight
Fusarium spp.	Brown foot rot, ear blight
Leptosphaeria nodorum	Glume blotch, leaf spot
Monographella nivalis	Snow mould, brown foot rot
Pyrenophora tritici-repentis	Yellow leaf spot, tan spot, leaf blight
Septoria nodorum	Glume blotch
Tilletia controversa	Dwarf bunt
Tilletia indica	Karnal bunt
Tilletia tritici	Bunt, rough spored bunt
Urocystis agropyri	Flag smut
Ustilago tritici	Loose smut
Bacillus megaterium pv. *cerealis*	White blotch
Clavibacter tritici	Yellow slime disease, yellow ear rot
Barley stripe mosaic virus	False stripe
Dilophosphora alopecuri	Twist, gall nematode
Anguina tritici	Ear cockle nematode

cut prior to the start of triticale's anthesis in order to reduce the risk of ergot (*Claviceps purpurea*) spreading to infect the seed crop.

Roguing

The roguing stages are as described for wheat.

Harvesting, drying and seed processing

These are as described for wheat. However there is a tendency for the seed head to break off at its neck in some cultivars. Some triticale cultivars produce seed that is larger than wheat. The seed moisture contents at harvest are as described for wheat.

Pests and pathogens

The main seed-borne pathogens of triticale are listed in Table 7.2.

Rice

- *Oryza sativa* L. – Asian Rice, Paddy Rice; and
- *Oryza glaberrima* Stend. – African rice.

Origins and types

The exact geographic origin of *O. sativa* is not certain: some authorities believe it to be China while others consider India as its cradle of origin. For detailed discussions and further background to the history and development of rice see Hancock (2004) and Chang (1984).

A genetically engineered rice, referred to as *Golden Rice*, has been developed; this new type has a grain high in vitamin A. The purpose of Golden Rice is to alleviate vitamin A deficiency in populations where rice is a staple food. Early work has introduced the character into the cultivars predominantly used in southern USA. The International Rice Research Institute (IRRI), based in the Philippines, is coordinating development of Asian types of Golden Rice for production in Asia (www.goldenrice.org).

Table 7.2. The main seed-borne pathogens of triticale; these may also be transmitted by other vectors.

Pathogen	Common name
Claviceps purpurea	Ergot
Cochliobolus sativus	Spot blotch, seedling blight, common root rot, head blight
Fusarium spp.	Head blight
Tilletia indica	Karnal bunt
Ustilago tritici	Loose smut
Clavibacter michigenensis ssp. *nebraskensis*	Goss' bacterial wilt

There are two rice types:

- Upland, which is cultivated in rain-fed conditions; and
- Irrigated, cultivated in a range of water environments.

Diversity of the *O. sativa* types has resulted from the gradual geographic spread of the Asian type evolving into the three subspecies:

- *Indica*; a short plant, mainly cultivated in the warm, humid tropics.
- *Japonica*; mainly day-neutral, some short-day cultivars; it can be cultivated outside the tropics.
- *Javonica*; day-neutral, mainly cultivated in the equatorial climate of Indonesia.

There are many diverse types within the three races.

Oryza glaberrima, African rice, is generally thought to have its primary source in the swampy areas of the upper Niger; however, this species has been largely superseded by types of *O. sativa*.

Classification of cultivars

For the classification of cultivars see UPOV Test Guideline 016 (see Appendix).

Site

Rice is usually cultivated continuously on the same site. Therefore special care should be taken when there is a change of cultivar or when producing a higher category seed class of the same cultivar. A change in cropping pattern should be made if there is a serious increase in the occurrence of wild rice, also known as 'red rice' (*Oryza rufipogon*), which is a serious weed of cultivated rice; wild rice can only be distinguished from the cultivars at maturity, and is also able to cross-fertilize with the cultivated types. Other important weed species frequently cited in seed standards for rice include: *Echinochloa* spp., *Panicum* spp., *Sorghum halepense* and *Cyperus rotundus*.

In rice production sites where water is only available for one rice crop the rice can be rotated with other crops which do not require flooding; several vegetable species can be grown as alternatives with significant economic advantage.

When the rice seed producer is changing to a new cultivar it is advisable to omit rice for 2 years; in some cases it is possible to use the first crop of replacement rice for ware or domestic purposes.

Pollination and isolation

Rice is generally self-fertilized. The isolation requirement for upland rice is 2 m when there is no other crop or physical barrier. Water-grown rice (swamp or floating) normally occupies an entire bund or enclosure, therefore isolation distance is not an issue. The recommended isolation distance for the female parent of hybrid rice seed production is 40 m from all other rice crops except the male parent.

Seed crop establishment

The soil pH can range from 5.0 to 6.5. The crop does not tolerate saline conditions.

The *O. sativa* subsp. *indica* cannot tolerate excess available nitrogen, although *O. s. japonica* can economically use higher nitrogen applications. Phosphorus levels should be maintained in order to achieve optimum plant growth; it can be applied in a water-soluble form for paddy rice production. The potassium available from either the soil or solution from the water source is usually sufficient.

Upland rice is established by similar methods as for other rain-fed small grain cereals; ideally drilling in rows at a sowing rate of 70–130 kg/ha.

Water grown rice is either sown direct or transplanted. The sowing rate is 70–130 kg/ha for direct sowing in a bund. Row sowing or planting out is especially helpful for roguing or seed crop inspection. The plant breeder should be consulted regarding sowing rates and the ratio of male to female for the production of hybrid rice seed.

Transplanting is more economical, especially in the earlier stages of multiplication when up to 25 kg of seed is necessary for a seedbed of 0.1 ha, which will provide sufficient transplants for 1 ha. Figure 7.1 shows a rice seedbed in Bangladesh with nursery plants being lifted for final planting out. The seedlings are subsequently planted 3–6 weeks from sowing. Figure 7.2 illustrates a transplanted plot. There are proprietary systems for producing rice seedlings in shallow water, using lightweight materials that provide physical support from sowing through to planting out. The depth of water, which is usually from 15 to 30 cm, should be maintained throughout flowering until the early stage of seed development. The retaining embankment or causeway (known locally as a 'bund') is breached when the seed is mature and the soil then dries out prior to harvest.

Roguing

Roguing should be in stages from the crop's first tillering through to the soft dough stage of seed maturity. Weed and wild rice control during roguing is an important additional task as also is removal of any rice plants showing disease symptoms. Roguing and weeding are more efficient when crops have been planted out.

Fig. 7.1. A rice seedling bed in Bangladesh, at transplanting stage.

Fig. 7.2. A transplanted rice plot for seed production in Bangladesh; note that the plants are in rows. There is a retaining bund in the middle ground.

Harvesting

The seed crop is usually ready for harvest 4–6 weeks following the end of anthesis. By this stage the seeds should be hard at the 'thumb nail' test, with seed moisture content at or slightly above 18%, but the seed moisture content should not be higher than 25% when the crop is cut.

Large areas of upland rice can be harvested by combines; when irrigated areas are combined so-called floatation tyres are fitted. Small plots can be cut by hand, the panicles tied in bundles and stooked or stacked for further drying; if the panicles are without straw they should be taken to a suitable drying floor where they can either be hand threshed by beating the rice panicles or the seed extracted by a small thresher.

Drying

Seed moisture content should be reduced to below 12% in humid climates. In the more arid areas a seed moisture content of 14% is satisfactory until the following sowing season. The seed moisture content should be reduced to 9% for vapour-proof storage.

Seed processing

Small quantities of seed are usually hand winnowed or dealt with on a small winnower. Rice seed is cleaned on air/screen cleaners. The screens used should take into account whether the seed to be processed is either short or long grained. Further upgrading can be achieved with an indent cylinder and/or a gravity separator.

Pests and pathogens

The main seed-borne pests and pathogens of rice are listed in Table 7.3.

Hordeum vulgare L. *sensu lato* (Barley)

Origins and types

Barley originates from southwest Asia; the crop's early history and domestication have been discussed by Harlan (1984a) and Hancock (2004). This species has a reputation

Table 7.3. The main seed-borne pests and pathogens of rice; these may also be transmitted by other vectors.

Pest or pathogen	Common name
Phoma spp.	Glume spot
Alternaria longissima	Alternaria
Alternaria padwickii	Stackburn disease, pink kernel, leaf spot, seedling blight
Balansia spp.	Incense rod, black ring, sterility disease, Udbatta
Cochliobolus miyabeanus	Seedling blight, brown spot, pecky rice, black sheath rod
Curvularia spp.	Kernel discoloration
Gibberella fujikuroi	Bakianae disease, foot rot
Monographella albescens	Leaf scald
Oospora oryzetorum	Blight
Pyricularia oryzae	Blast, rotten neck
Pseudomonas avenae	Brown stripe
Xanthomonas campestris pv. *oryzae*	Bacterial leaf blight
Aphelenchiodes besseyi	White tip (a seed-borne nematode)
Ditylenchus angustus	Ufra, dak-pora, rice stem nematode

for succeeding where other cereals are not so prolific. It is a cool-season crop, tolerant of high temperatures but requires a low relative humidity. The main uses are stock feed and malting although some grain is also used locally for human consumption. There are spring and winter types although both types can be sown in spring.

For the classification of cultivars see UPOV Test Guideline 019 (see Appendix).

Site

Sites proposed for barley seed production should not have grown it for the previous 2 years, as the seed is able to survive in the soil for several years, except when the same cultivar is to be grown from authentic seed. The six-row cultivars are especially persistent as dormant seeds. In tropical areas barley seed production can follow in rotation with pulses or root crops; in drier areas of the tropics it can follow a fallow period. The field border areas of barley crops should be kept cultivated in order to minimize the spread of pests and pathogens.

Pollination and isolation

Barley is generally considered to be self-fertilized, although some of the six-row winter cultivars can cross-pollinate. When cross-pollination is likely to occur isolation distances should be 300 m, otherwise 2 m is sufficient when there is no physical barrier. A non-cereal barrier crop provides a useful protection against the control of some air-borne pathogens, such as smut, with a minimum isolation distance of 500 m.

Roguing

Straw-shortening chemicals should not be used as they are likely to conceal morphological characters. The optimum stage for roguing is immediately following ear emergence; this is when the seed has started to fill.

Harvesting

The seed should be just hardened and difficult to mark with a thumb nail. At this stage the straw will have lost its green pigmentation. Seed should not be harvested until its moisture content has reduced to 20% or less otherwise the seed will be mechanically damaged. Seeds which have a moisture content of 8% or less are also similarly vulnerable. It is normal practice to attach a de-awner to the combine or thresher. Large areas can be combined.

Drying

When seed is destined for sowing in the following sowing season it is stored at 14% moisture content. If stored for longer periods of approximately 1 year the moisture content is reduced to below 10%; seed to be stored for longer periods should be reduced to below 8%. Air drying temperature should not exceed 49°C; when drying from a greater moisture content the air drying temperature should be reduced to 44°C.

Seed processing

The air-screen cleaner provides the best initial upgrading; coupling in sequence with indent cylinders and a gravity separator can improve the purity of the seed lot. A de-awner is used prior to upgrading the seed lot if an appropriate attachment was not included while combining or threshing.

Pests and pathogens

The main seed-borne pathogens of barley are listed in Table 7.4.

Avena spp. (Oats)

There are three *Avena* spp. in cultivation, these are:

- *A. sativa* L. – oat;
- *A. byzantina* K. Koch. – the naked oat; and
- *A. byzantina* K. Koch. – the red oat.

Avena sativa is the oat crop dealt with in this volume. The naked oat is little grown, mainly because it is low yielding. The red oat is cultivated in the Mediterranean area, Australia and South Africa. There are hybrid cultivars resulting from crossings of *A. sativa* and *A. byzantina*. Seed of both the naked oat and the cultivated red oat are produced in the same way as described below for oat.

Origins and types

The *Avena* species originated in the Middle East, the Mediterranean basin and Ethiopia, although it was not established as a separate crop in central Europe until approximately 3000 years ago (Hancock, 2004). The evolution and history of more recent domestication of oats has been discussed in detail by Holden (1984).

For the classification of cultivars see UPOV Test Guideline 020 (see Appendix).

Site

Ideally the proposed seed multiplication site should not have grown any oat crop for a minimum of 2 years; national seed regulations

Table 7.4. The main seed-borne pathogens of barley; these may also be transmitted by other vectors.

Pathogen	Common name
Alternaria spp.	Black point, kernel smudge
Claviceps purpurea	Ergot
Cochliobolus sativus	Seedling blight, foot rot, common root rot, black point
Gibberella zeae	Scab, seedling blight, foot rot
Monographella nivalis	Snow mould, brown foot rot
Pyrenophora graminea	Leaf stripe
Pyrenophora teres	Net blotch
Rhynchosporium secalis	Scald
Ustilago avenae	Black smut, false loose smut, semi-loose smut
Ustilago segetum var. *hordei*	Covered smut
Ustilago tritici	Loose smut
Barley mosaic virus	False stripe

may be more specific on this requirement, depending on the seed class to be produced. Seeds of wild oat species, which include *A. fatua*, *A. sterilis* and *A. ludoviciana*, are virtually impossible to remove from oat seed lots. Therefore every effort should be made to ensure a good control of wild oat on the proposed site in the years before the seed crop. Experience has shown that it can take up to 4 years to eradicate dormant wild oat seed from a site. Other difficult weeds include: *Raphanus raphanistrum*, *Agrostemma githago* and *Lolium temulentum*.

Pollination and isolation

This species is self-fertilized, therefore either a physical barrier or a minimum crop gap of 2 m is normally all that is required.

Roguing

There are two possible roguing stages:

1. Immediately following panicle emergence, ideally just as the grain is starting to fill.
2. When seed (lemma) colour changes can be observed.

Harvesting

Oat seeds tend to ripen less evenly in the panicle than many other cereals although many cultivars in current crop production can be efficiently harvested by combine. The oat seed should be hard and difficult to mark with a thumb nail; the straw tends to retain its green pigmentation at seed maturity more than other cereals.

Drying

The air drying temperature should commence at approximately 44°C, with an increase to 49°C as the drying proceeds. The oat seed lot's moisture content should reduce down to at least 16% for very short-term storage. For storage over several months moisture content should be reduced to 14% or less. The optimum seed moisture content is 10% when long-term storage is visualized; this should be reduced to 8% if the seed is to be stored in vapour-proof containers.

Seed processing

The seed can normally be cleaned successfully with an air-screen cleaner. Passing the seed lot through an indent cylinder will improve the seed lot's physical quality. Some seed processors 'clip and polish' the seed; although this improves the seed appearance and mechanical flow it may adversely affect germination.

Pests and pathogens

The main seed-borne pests and pathogens of oats and other *Avena* spp. are listed in Table 7.5.

Table 7.5. The main seed-borne pests and pathogens of oats and other *Avena* spp; these may also be transmitted by other vectors.

Pest or pathogen	Common name
Alternaria spp.	Smudge, black point
Claviceps purpurea	Ergot
Fusarium spp.	Brown foot rot, ear blight
Pseudomonas syringae pv. *striafaciens*	Bacterial stripe blight
Pseudomonas syringae pv. *coronofaciens*	Halo blight
Pyrenophora avenae	Leaf (stripe) spot
Ustilago hordei	Covered smut
BaYMV-M and BaYMV	Barley stripe mosaic virus
Anguina tritici	Spikelet nematode
Ditylenchus dipsaci	Tulip root, segging

Further Reading

General

FAO (2002) The challenges after Rio. Available at: www.faoorg/ag/magazine/0102sp1.htm (Accessed 2 September 2011).

Hancock, J.F. (2004) *Plant Evolution and the Origin of Crop Species*, 2nd edn. CAB International, Wallingford, UK.

Wheat

FAO (2002) *Bread Wheat*. Plant Production and Protection Series No. 30, FAO, Rome.

Edwards, I.B. (1987) Hybrid varieties of wheat. In: Feistritzer, W.P. and Kelly, A.F. (eds) *Hybrid Production of Selected Cereal Oil and Vegetable Crops*. FAO, Rome, pp.3–25.

Stroike, J.E. (1987) Technical and economic aspects of hybrid wheat seed production. In: Feistritzer, W.P. and Kelly, A.F. (eds) *Hybrid Production of Selected Cereal Oil and Vegetable Crops*. FAO, Rome, pp.177–185.

Rice

Virmani, S.S. (1987) Hybrid rice breeding. In: Feistritzer, W.P. and Kelly, A.F. (eds) *Hybrid Production of Selected Cereal Oil and Vegetable Crops*. FAO, Rome, pp.35–54.

Xizhi, L. (1987) Technical aspects of seed production of hybrid varieties of rice in China. In: Feistritzer, W.P. and Kelly, A.F. (eds) *Hybrid Production of Selected Cereal Oil and Vegetable Crops*. FAO, Rome, pp.187–192.

8

Gramineae – Cross-Fertilized Cereals

Rye, Maize, Sorghum and Millets

The important cross-fertilized cereals are:

- *Secale cereale* L. – rye;
- *Zea mays* L. – maize;
- *Sorghum bicolor* (L.) Moench – sorghum;
- *Pennisetum glaucum* (L.) B. Br. Emend. syn. *P. americanum* (L) Lecke, also syn. *P. typhoides* (Burm.) Staph et Hubb. – pearl millet or bulrush millet; and
- *Eleusine coracana* (L.) – finger millet.

Secale cereale L. (Rye)

Origins and types

Rye is generally regarded as having originated in southwest Asia. This species produces an economic crop in those geographical areas with cold continental winters and arid summers as experienced in eastern Europe and parts of North America (Evans, 1984).

The crop is used for the production of rye bread, biscuits and crisp-breads; the North American crop is mainly used for the production of rye whisky. Some rye crops are cultivated specifically for fodder, stock-feed or the production of rye starch.

For the classification of cultivars see UPOV Test Guideline 058 (see Appendix).

Site

The proposed site should be free of wild oat species and as far as possible other undesirable weeds including *Raphanus raphanistrum* (wild radish), *Agrostemma githago* (corncockle) and *Lolium temulentum* (darnel). Precautions should be taken to minimize the risk of ergot infection in the seed crop. Rye is tolerant of slightly acid and also sandy soils, although the best yields are obtained when grown in more fertile conditions. Nutrient requirements are as for barley.

Pollination and isolation

Rye is cross-pollinated and should have an isolation distance of 250 m and a 2-year break from other cereal species. This cereal species sheds readily and dormant seeds from previous crops can present difficulties regarding cultivar purity. However, rye can follow rye if both crops are the same cultivar produced from authentic seed. Crops of diploid and tetraploid cultivars should also be isolated from each other because pollination by parents of a different ploidy can result in infertility.

Seed crop establishment

The rye crop is sown at the rate of 80–150 kg/ha. Suitable walkways should be left for roguing or crop inspection. The recommended ratios of parents for the production of hybrid cultivars should be obtained from the breeder; the appropriate proportions of seed should be received for the production of synthetic cultivars.

Roguing

The seed crop should be rogued immediately following ear emergence as this is the optimum stage to check the plants' morphological characters. Additionally all plants that display symptoms of ergot should be removed during roguing.

Harvesting

Rye seed should be harvested when there is difficulty in marking the seed with the thumb nail; at this stage the straw's green pigmentation will have faded. Threshing should not commence until the seed moisture content is down to 20% or slightly lower; if threshed at higher moisture levels there is a risk of seed damage resulting in reduced germination.

Drying

The drying process for rye should be done slowly. Where local climate permits, the seed can be spread on a clean and smooth surface, frequently turned and dried in the open air. The air temperature for forced-air drying should not exceed 44°C when the seeds' moisture content is above 16%. When the seed moisture content is lower the air temperature can be 49°C. Seed moisture content should be immediately reduced to 16% or slightly lower following harvest by combine. The safe level for seed storage until the following sowing season is 14%; for longer term storage it should be reduced to 8%.

Seed processing

Rye seed is extremely susceptible to mechanical damage during processing. The most efficient method of cleaning rye seed is by an air/screen cleaner. When necessary the quality of the seed lot may be further enhanced by passing the seed through an indent cylinder and/or a gravity separator. If galls of *Anguina* are observed in the seed lot they should be removed by gravity table during the course of processing; the galls may also contain *Dilophospora alopecuri*, which is responsible for 'twist' or 'plumed spore disease'. In addition, sclerotia of *Botrytis* and *Claviceps* can be removed by a gravity table during seed processing.

Pests and pathogens

The pests and pathogens of rye are listed in Table 8.1.

Zea mays L. (Maize)

Origins and types

Maize originates from Central and South America. According to Hancock (2004) the history of maize has been difficult to trace as its fruiting cob and its monoecious nature are unique in the *Gramineae*. The hypotheses regarding maize evolution and its subsequent development as a food and fodder crop have been further discussed by Hancock (2004).

According to Purseglove (1985), the cultivars of *Zea mays* can be classified into seven groups:

1. Pod corn – *Zea mays tunica* Sturt.
2. Popcorn – *Zea mays everata* Sturt. (syn. *praecox*).
3. Flint maize – *Zea mays indurata* Sturt.
4. Dent maize – *Zea mays indenata* Sturt.
5. Soft or flour maize – *Zea mays amylacea* Sturt. (syn. *erythrolepis*).
6. Sweetcorn – *Zea mays saccharata* Sturt. (syn. *sugosa* Bonof.).
7. Waxy maize – *Zea mays ceritina* Kulesh.

Table 8.1. Pests and pathogens of rye; these pathogens may also be transmitted to the crop by other vectors.

Pest or pathogen	Common name
Claviceps purpurea	Ergot
Fusarium spp.	Foot rot
Gloeotinia granigena	Blind seed disease
Monographella nivalis	Snow mould, brown foot rot
Tilletia controversa	Dwarf bunt
Tilletia tritici	Rough-spored bunt
Tilletia laevis	Smooth-spored bunt
Urocystis occula	Stem smut, stalk smut
Xanthomonas campestris pv. *undulosa*	Black chaff
Anguina tritici	Eelworm galls, ear cockle

The different cultivars of these groups cross-pollinate and it is sometimes difficult to classify them exactly. The types mainly used in agriculture are flint, dent and waxy, although the flour type is cultivated in some local areas.

Classification of cultivars

For the classification of cultivars see UPOV Test Guideline 022 (see Appendix).

Site

Dormant maize or weed seeds are not normally a problem with seed production sites. However, maize seedlings do not compete well with weeds, and it is therefore important that the young plants be kept weed free in the early stages of crop establishment. Maize tolerates a range of soil types, but it is important that the soil should have satisfactory water retention characters. The crop tolerates a soil pH range of 5.5–6.8; appropriate quantities of lime should be applied if the pH is below this. The N:P:K fertilizer application during the final stages of soil preparation should be in the ratio of 1:2:2. Some growers apply a higher proportion of nitrogen in the base dressing while others give a nitrogenous top dressing during the young plant stage if a lack of vigour is observed.

Seed crop establishment

Maize seed is generally sown in rows 50–100 cm apart; there is a tendency to adopt the wider row spacing in the tropics. When the required row spacing is 75 cm and the target population is 50,000/ha, the distance from station to station within the rows is 26 cm. The selected plant density depends on the growing conditions; 50,000/ha is the optimum. Information regarding optimum spacing and row patterns and parent ratios for hybrid cultivar production should be obtained from the maintenance breeder. Care should also be taken to ensure that the hybrid's parents reach anthesis at the same time; local experience should provide evidence to assist decisions regarding sowing the parental material.

Pollination and isolation

The male flowers ('tassels') are in a terminal inflorescence and the female flowers ('silks') are borne in conjunction with the 'cobs' or 'ears' on lateral branches. There is a tendency for dehiscence of pollen up to 2–3 days in advance of the stigmas becoming receptive, which increases the incidence of cross-pollination.

Maize pollen is wind-borne and distances of up to 1 km may be required by some seed regulatory authorities to isolate seed crops of sweetcorn from other types of *Zea mays*. The usual isolation distance is a minimum of 200 m with greater distances of up to 600 m for the higher seed class. In the case of open-pollinated cultivars the isolation distance can be reduced if the seed from the outer rows is collected separately and omitted from the main seed lot; the number of rows to be discarded depends on the field size and the isolation distance; a table for determining this is given in Kelly and George (1998c). Some seed schemes allow a reduced isolation distance for hybrid seed production when the borders or perimeters of fields are planted with the male parent.

Roguing

Special attention should be given to the removal of any volunteer plants of *Zea mays* from seed production fields before the start of anthesis.

Both parents must be inspected if hybrid seed is being produced and roguing criteria applied according to the description of each parent line. The roguing operations must be completed before male and female inflorescences are fully developed.

An additional check for trueness to type (especially characters such as seed colour) can be made after harvesting when the husks are removed from the cobs.

Hybrid maize seed production

Parental inbred lines for hybrid seed production are normally maintained by, or under the supervision of, the responsible maintenance breeder. Advice on ratio of female to male parent lines should be sought from the breeder in order to ensure adequate pollination and maximum yield of hybrid seed. The usual ratio of female to male lines is 4:2, although 4:1 may be adequate for the production of some hybrids. The relative height of the male parent can have some influence on this ratio because pollen distribution is more effective when the male parent is taller than the female.

Unless there is satisfactory genetic or chemical control of pollen production, the male inflorescences ('tassels') are removed from the female parent line. This operation, referred to as 'de-tasselling', must be done before the anthers dehisce; it may also be necessary to remove male inflorescences from the tillers of the female parent. Various methods for de-tasselling are used, including cutting by hand, working from mobile platforms or using machines with high-level rotating cutters. It may be necessary to de-tassel a crop several times to ensure that self-pollination of the female parent does not take place. Whichever system is used it is important to ensure that the male inflorescences of the female parent are removed completely before they shed pollen.

Harvesting, drying and processing

When hybrid seed is being produced it is important to brief field workers on the layout and ratio of female to male lines before any harvesting commences. Clear instructions must be given to harvesting gangs regarding the disposal of mature cobs from the male lines.

The seed becomes hard prior to harvesting stage; another indication of maturity is the yellowing of the mother plants' foliage and main stems. If the seed crop is to be mechanically harvested it is best done with a corn-picker. This machine gathers the ears, and dehusks them.

The cobs of the seed crop are mechanically harvested when the seeds' moisture content is down to about 25% or slightly less. For hand harvesting, a moisture content of 15–20% is more suitable. Visual signs are when the plants start to become yellow and the seeds have hardened. Small areas are harvested by hand but specialists in the USA either use mechanical pickers or mechanical combining; care must be taken with the use of mechanical combines when producing seed crops as distinguished from ware crops in order to avoid significant damage to the harvested seeds. The husks are removed from the cobs immediately after harvesting leaving the 'ears' with the seeds attached. In the early stages of multiplication of seed stock, examination of the ears provides an ideal opportunity to reject material which is either diseased or displaying off-type characters. Further drying of the ears is either done in bird- and rodent-proof mesh cages known as 'cribs' or they are dried artificially. The air temperature for artificial drying should not exceed 42°C. The drying process should commence at a slow rate and there should be an efficient air circulation through the ears; therefore depth of ears should not be excessive otherwise the throughput of warm air will be restricted.

Once the maize ears have dried down to 14% moisture content they can be shelled. Depending on the scale of production and the availability of hand labour the ears are either shelled by hand or machine; this operation is known as 'shelling', which is the removal of seeds from their cobs. Modern shelling machines are designed to minimize damage to the seeds; a typical maize shelling compound is illustrated in Fig. 8.1.

Following shelling the moisture content of the seed lot should be determined. For short-term storage it should not exceed 14% moisture content; the moisture content for long-term seed stocks should not exceed 10%.

The shelled seeds may require separation from pieces of husk and other cob debris before being passed through an air-screen cleaner. The cleaned seed lot is usually graded in order to separate the different sized and shaped seeds originating from different parts of the cob; this can be achieved by upgrading the seed lot on a gravity separator or an indent cylinder.

Small-scale farmer's maize cob storage

Many small-scale and subsistence farmers have to save their own open-pollinated maize seed for production of the following year's crop. One way to achieve this is to store the cobs over the kitchen stove. This has a double advantage of deterring seed storage weevils and also storing the seeds in a low relative humidity. The seed are shelled from the cob just before sowing time, graded and the healthier looking seeds sown; see Fig. 8.2.

Pests and pathogens

The main seed-borne pathogens of *Zea mays* are listed in Table 8.2.

Fig. 8.1. A maize sheller with crib-dried maize cobs ready for shelling.

Fig. 8.2. Kitchen storage of maize cobs for producing the next season's crop. Photograph by kind permission of The Real IPM Company (K) Ltd (www.realipm.com).

Table 8.2. The main seed-borne pathogens of *Zea mays*; these pathogens may also be transmitted to the crop by other vectors.

Pathogen	Common names
Acremonium strictum	Kernel rot
Claviceps gigantea	Ergot
Cochliobolus carbonum	Charred ear mould, southern leaf spot
Cochliobolus heterostrophus	Southern leaf spot, blight
Colletotrichum graminicola	Anthracnose
Diplodia frumenti	Dry ear rot, stalk rot, seedling blight
Fusarium spp.	Fusarium
Gibberella fujikuroi	Gibberella ear rot, kernel rot, stalk rot
Gibberella fujikuroi var. *subglutinans*	Seedling blight
Gibberella zeae	Seedling blight, cob rot
Khuskia oryzae	Nigrospora cob and stalk rot, dry rot
Marasmius graminum	Seedling and foot rot
Penicillium spp.	Seed rot, blue-eye
Peronosclerospora sacchari	Sugar cane downy mildew
Peronosclerospora sorghi	Sorghum downy mildew
Sclerophthora macrospora	Crazy top
Stenocarpella macrospora and *S. maydis*	Dry or white ear rot, stalk rot, seedling blight, root rot
Ustilaginoidea virens	False smut, green smut
Ustilago zeae	Smut, blister smut, loose-smut
Erwinia stewartii	Bacterial leaf blight, bacterial wilt, Stewart's disease, white bacteriosis
Maize leaf spot virus	
Maize mosaic virus	
Sugar cane mosaic virus	

Sorghum bicolor (L.) Moench (Sorghum)

Origins and types

The cultivated *S. bicolor* subsp. *bicolor* was probably derived from *S. bicolor* subsp. *arundinaceum* following human selection for non-shattering heads, large seeds and heads, threshability and suitable maturities (Doggett and Rao, 1995). The evolution and possible origins of sorghum have been discussed by Hancock (2004). Sorghum is an important crop in areas which are not suited to maize, because sorghum can tolerate and be successful in drier conditions. Although sorghum can be produced from a ratoon crop, this is not normally recommended for the multiplication of seed stocks, although it can be useful for the production of ware crops.

For the classification of cultivars see UPOV Test Guideline 122 (see Appendix).

Site

Seed production areas and sites subject to strong winds are not suitable for the taller cultivars. Sorghum is often regarded as a demanding crop for soil nutrients, therefore seed crops are usually rotated with a previous leguminous crop.

The main weed problems in sorghum seed crops arise from the presence of *Striga* spp., wild sorghum spp., volunteer forage sorghums and *Sorghum halepense* (Johnson grass), which is an important weed worldwide and is a reservoir of infection, especially for the sorghums. Figure 8.3 shows sorghum infected by *Striga*. Aleppo sorghum (*Sorghum halepense*) can be a significant weed in a hybrid sorghum crop. Bearing in mind the various serious problems arising from these weeds and volunteers, sites which are proposed for the production of grain sorghum seed should ideally be clear of their propagules and dormant seeds. Sorghum produces satisfactory seed crops on soils with a pH range of 5.5–7.5.

Pollination and isolation

Sorghum is generally regarded as being self-pollinated, however it will out-pollinate for up to 10%; the minimum isolation distance is 200 m but 400 m should be allowed between different cultivar types.

Seed crop establishment

Inter-row spacing is dependent on the tillering ability of the cultivar being multiplied. The non-tillering cultivars are produced in rows, usually 15–30 cm apart. The inter-row distance for the tillering cultivars is 45–90 cm. The seed sowing rate for the narrow spaced cultivars is 10–20 kg/ha, while the sowing rate is 6–10 kg/ha for cultivars grown at wider spacings.

Weed control is important, especially in the early stages of crop establishment; however, care should be taken that cultivations should be shallow to avoid root damage.

Roguing

Open-pollinated crops are usually rogued two to three times and hybrid cultivars three times. Any plant not conforming to the material's morphology should be removed as soon as it is observed. It is important that both open-pollinated and hybrid cultivars are checked to ensure that isolation distances are according to requirements and also that no female plants in the hybrid cultivars are shedding pollen. The crops are again rogued to confirm cultivar purity according to seed colour.

Harvesting and drying

The correct harvesting stage is indicated by firmness of the seed, which cannot be dented with the thumb nail; the seed will usually have a moisture content of below 15%. A good indication is that by this stage the lower leaves of the mother plants will

Fig. 8.3. Sorghum infected with the plant parasite *Striga*. Photograph by kind permission of The Real IPM Company (K) Ltd (www.realipm.com).

have become detached and also that in many cultivars the upper leaves have become yellow. Maturing sorghum seed can start to germinate in high relative humidity; it is therefore prudent to commence harvesting in inclement weather or prior to predicted rainfall.

Small-scale areas can be harvested by hand, using knives or small sickles. Large-scale areas may be harvested by combine; this is normally used for the short-straw and uniform cultivars. Small quantities of harvested seed heads can either be further dried in the field or transferred to a clean drying floor. The depth of drying seed heads should not be greater than 20 cm; the material being dried must be turned at frequent intervals. Seed lots of combined seed should be dried using forced air at a temperature of 40°C or lower. The final seed moisture content for open storage is 10% or less. For storage in vapour-proof containers the moisture content should be further reduced to 9%.

When harvesting the hybrid cultivars it is important to brief workers as to the row patterns and ratios of male and female parental material. The seed heads of the male parent should be harvested and removed from the site before attempting to harvest the hybrid seed crop from the female parent plants.

Seed processing

The initial pre-cleaning of sorghum seed lots is either completed by winnowing while on the threshing floor or by a pre-cleaner. Further upgrading is completed with an air/screen cleaner.

Pests and pathogens

The main seed-borne pests and pathogens of sorghum are listed in Table 8.3.

The Millets

Pennisetum glaucum (L.) B. Br. Emend. syn. *P. americanum* (L.) Lecke, syn. *P. typhoides* (Burm.) Staph et Hubb. (Bulrush Millet, Pearl Millet, Spiked Millet, Cat-tail Millet, Bajra)

Origins and types

Bulrush millet is thought to have its centre of origin in the African highlands and was subsequently taken via East Africa to the Indian subcontinent in the second millennium BC (Purseglove, 1984).

For the classification of cultivars see UPOV Test Guideline 260 (see Appendix).

Site

Pennisetum glaucum grows successfully in the more arid areas of the tropics. The crop does not tolerate heavy rainfall, which can affect pollination efficiency. This millet responds to nitrogen at the rate of 120 kg/ha that has been incorporated during site preparation.

The crop can be usefully rotated with groundnuts or other legumes of local importance in tropical areas; it should not be grown after any of the millets or other small-seed crop species. The millet seed crop is frequently followed by sorghum or cotton. There should always be at least a 1-year break between millet crops. As with sorghum the millets are prone to parasitism by *Striga* spp.

Pollination and isolation

Pennisetum glaucum is mainly cross-pollinated as a result of having two types of inflorescence; the first flowers are hermaphrodite, these are closely followed by the anthesis

Table 8.3. Seed-borne pests and pathogens of sorghum; these may also be transmitted to the crop by other vectors.

Pest or pathogen	Common name
Bird species, including parrots	
Ascochyta sorghi	Rough leaf spot disease
Claviceps spp.	Ergot, asali, sugary disease
Cochliobolus sativus	Seedling blight
Colletotrichum graminicola	Stalk rot, red leaf
Fusarium spp.	Fusarium
Glomerella tucamanensis	Red rot, midrib spot
Mycosphaerella holci	Red leaf, Phyllosticta seed disease
Periconia circinata	Root rot, crown rot, milo disease
Peronosclerospora sorghi	Downy mildew, leaf shredding disease
Sphacelotheca reiliana	Head smut
Sphacelotheca cruenta	Loose smut
Sphacelotheca sorghi	Covered smut, grain smut
Pseudomonas syringae	Bacterial leaf spot
Xanthomonas rubrisorghi	Bacterial leaf streak

of male flowers, i.e. with stamens only, thus ensuring an efficient system of predominant cross-pollination. Isolation distances should be at least 300m, with up to 1000m for the higher seed classes. Some cultivars have very significant defining morphological characters and should have an increase in isolation distance according to the seed class being produced. This requirement is particularly important for isolation of the grain cultivars from the fodder types (the fodder millets are discussed in Chapter 9.

Seed crop establishment

The crop requires a soil pH of 5.5–6.5. It is tolerant of dry soils although when the soil moisture content is replenished it does not have the post-drought recovery potential of sorghum. The application of nitrogenous fertilizers during site preparation will generally significantly increase seed yields.

The inter-row distances depend on whether or not the crop will be rain-fed or irrigated. When sown on the flat, row distances are 50cm apart with a sowing rate of 6–8kg/ha. When sown on ridges the rows are 90cm apart with a sowing rate of 6kg/ha. The crop should be kept weed free, especially while it is establishing.

For the production of hybrid seed the ratio of female to male parents should be according to the maintenance breeder's specifications, according to the hybrid cultivar being produced. The outer rows of the seed production plot or field should be four rows of the male parent.

Roguing

The isolation of the seed crop should be checked prior to anthesis. The next stage is to check cultivar characters. In hybrid cultivar seed production the checks are required to ensure that no female plant is shedding pollen.

Harvesting

Bulrush millet is best harvested by hand, especially as some types produce significant amounts of seed from secondary tillers thus making the total harvest very protracted. The remaining ear-head stems are usually used as a secondary straw crop. Hand harvesting can commence when the seed moisture content is approximately 25%; however, if harvest conditions permit it is preferable to wait until the seed moisture content has reduced down to 15%. Bulrush millet tends to germinate in the ear. The larger crop areas can be harvested by combine.

The harvesting of hybrid cultivars should commence with removal of the male parent rows. Field staff should be clearly informed as to the row pattern. Seed is then harvested from the remaining female parent rows.

Drying

Prior to threshing the seed heads should be dried down to 15% moisture content. The drying method used depends on local conditions and quantity to be dried: drying floors with the depth of seed not exceeding 20cm; sun-drying or with forced air, with its air temperature not exceeding 40°C.

Seed processing

Following threshing the separated seed should be either winnowed or passed through a pre-cleaner, the choice usually depending on the scale of production; small quantities are usually winnowed by hand, especially where there is adequate available labour.

Pests and pathogens

The pests and pathogens of the millets are listed in Table 8.4.

Eleusine coracana (L.) (Finger Millet, African Millet, Koracan, Ragi)

Origins and types

This millet is thought to have originated in Uganda and its neighbouring areas and is

Table 8.4. Pests and pathogens of the millets; the pathogens may also be transmitted to the crop by other vectors.

Pest or pathogen	Common name
Bird spp.	
Claviceps spp.	Ergot
Cochliobolus lunatus	Ear blight, seed rot
Cochliobolus setariae	Leaf blight
Fusarium spp.	Head mould
Gloeocercospora sorghi	Zonate leaf spot
Sclerospora graminicola	Green ear
Tolyposporium penicillariae	Smut
Pseudomonas syringae	Bacterial spot of *P. purpureum* (elephant grass)

believed to have been introduced into secondary centres in India from this original centre (Purseglove, 1984). The modern forms of finger millets are of major importance for food security in Ethiopia, Uganda and some parts of the Indian subcontinent. The species has small, reddish-brown coloured seeds; there are also white-seeded types of this species.

Mehra (quoted in Purseglove 1985) describes two groups of the finger millets:

1. African highland types with long spikelets, long glumes, long lemmas and with grains enclosed within the florets.
2. Afro-Asiatic types with short lemmas and with mature grains exposed in the florets.

The main distinguishing characters used for cultivar descriptions are:

- plant height: dwarf, tall;
- colour of vegetative parts: green, purple;
- tillering: little, much;
- type of inflorescence: spikes straight and open; spikes incurved and closed; spikes branched resembling a cockscomb;
- length of spikes: short, long;
- number of spikelets per spike: few, many;
- length of spikelet: short, long;
- tightness of packing of grain: lax, dense;
- length of glumes: short, long; and
- seed colour: white, orange red, deep brown, purple/black.

Site

The finger millets will yield satisfactorily on soils with a pH ranging from 5.0 to 6.5. The crop requires adequate soil moisture but cannot tolerate waterlogged soil conditions. Many seed producers rotate the crop with legumes; it should not immediately follow other small-grain cereals.

The cultivated crop readily cross-pollinates with the wild millet *Eleusine africana;* although morphologically similar to finger millet this weed is objectionable. Care must be taken to ensure that the proposed site and its surroundings are kept free of wild and other millets. The plant parasites *Striga* spp. can also be a problem with this crop.

Pollination and isolation

The crop is predominantly self-pollinated although up to 1% cross-pollination can occur. It is therefore necessary to ensure an isolation distance of 50 m from compatible pollen sources.

The seed crop establishment, harvesting and processing are as described above for bulrush millet.

Storage

Finger millets can be stored for several years as a grain crop without deterioration or significant weevil damage, thus providing an excellent species for food security in less favourable situations.

Pests and pathogens

The main seed-borne pests and pathogens of the millets are listed in Table 8.4.

Further Reading

Maize

Feistritzer, W.P. (ed.) (1982) *Technical Guideline for Maize Seed Technology*. FAO, Rome.

Kim, S.K. (1987) Hybrid varieties of maize. In: Feistritzer, W.P. and Kelly, A.F. (eds) *Hybrid Production of Selected Cereal Oil and Vegetable Crops*. FAO, Rome, pp.55–81.

Sorghum and millets

Chopra, K.R. (1987) Technical and economic aspects of hybrid varieties of sorghum. In: Feistritzer, W.P. and Kelly, A.F. (eds) *Hybrid Production of selected Cereal Oil and Vegetable Crops*. FAO, Rome, pp.193–216.

Doggett, H. (1988) *Sorghum*. Longman, London.

Matlan, P. (1990) Improving productivity in sorghum and pearl millet in semi-arid Africa. *Food Research Institute Studies* 22, pp.1–44.

Patil, T.T. (1987) Technical and economic aspects of seed production of hybrid varieties of millets. In: Feistritzer, W.P. and Kelly, A.F. (eds) *Hybrid Production of Selected Cereal Oil and Vegetable Crops*. FAO, Rome, pp.217–252.

9

Gramineae – Temperate, Prairie and Tropical Grasses

The first group of grasses dealt with in this chapter is the temperate species; these are followed by the prairie grasses, which are mainly produced for areas with a continental climate of predominantly cold winters followed by hot summers. The final part of this chapter deals with those grass species that are either sown or naturally occurring in the humid or drier tropical and subtropical areas.

The Temperate Grasses

The cytotaxonomic background and the early and recent histories of the temperate grasses have been discussed by Borrill (1976), who describes these backgrounds for *Lolium*, *Festuca*, *Dactylis*, *Phleum* and *Bromus* species.

Lolium spp. – Ryegrasses

The two *Lolium* spp. dealt with here are:

- *Lolium multiflorum* Lam. – Italian ryegrass, Westerwold; and
- *L. perenne* L. – perennial ryegrass, ray grass, English ryegrass.

Origins and types

This has been the most popular ryegrass in recent history; it was known to have been cultivated with irrigation in Lombardy and the Piedmont area of Italy since the end of the 12th century. The species was subsequently distributed to other areas of Western Europe (Borrill, 1976).

There are some annual cultivars, generally referred to as Westerwolds; other cultivars are biennials.

The main characters for the identification of ryegrass cultivars are:

- emergence of inflorescence: early or late;
- plant habit when inflorescence emergences: erect or prostrate;
- flag leaf length at emergence of inflorescence: short or long;
- flag leaf width at emergence of inflorescence: narrow or wide;
- stem length, to include fully developed inflorescence: short or long;
- leaf colour in autumn: light or dark green;
- plant morphology in spring of second year: erect or prostrate; and
- relative plant height in spring of second year: short or long.

There are some cultivars which have resistance to 'blind seed disease'; the causal pathogen of this disease is seed-borne and it infects the grass caryopsis tissue resulting in mummification.

For the classification of cultivars see UPOV Test Guideline 004 (see Appendix).

Site

Grass species, especially those with seed of similar size to those of the crop species to be produced, can present major weed competition and also seed processing difficulties. It is therefore generally recommended that grass species should not have been cultivated on the site for at least 2 years. The notable grass species to be controlled include: *Agropyron repens* (couch-grass), *Alepecurus myosuroides* (black grass), *Avena fatua* and *A. ludoviciana* (wild oats), *Bromus* spp. (bromes), *Holcus lanatus* (Yorkshire fog), *Lolium temulentum* (darnel), *Poa annua* (annual meadow grass) and *P. trivialis* (rough stalked meadow grass). The other weed species which need to have been controlled include cleavers and chickweed.

A soil pH of 6.0 is very satisfactory; when the pH is less than 5.0 liming material should be added during preparations. The soil should have satisfactory levels of available phosphorus and potassium; their levels should be checked and increased where indicated during soil preparations.

Pollination and isolation

Lolium multiflorum is cross-pollinated resulting from wind dispersal. *Lolium multiflorum* and *L. perenne* should be regarded as the same species from the point of view of isolation. The recommended isolation distance for fields of 2 ha or less is 200 m but 100 m for fields larger than 2 ha if the seed is intended for further multiplication. These recommended distances may be reduced to 100 m and 50 m, respectively, if the seed crop is only intended for fodder or other use.

Diploid and tetraploid cultivars do not cross-fertilize although the pollen may result in infertility thereby reducing potential seed yield; therefore it is generally recommended that the two types have an isolation distance of at least 50 m.

Seed crop establishment

The crop should be sown at a uniform depth, therefore drilling in rows from 10 to 20 cm apart is preferable to broadcasting unless the prepared seedbed is considered too wet to drill successfully.

An initial nitrogen application should be given to autumn-sown young plants to encourage tillering before the start of winter; similarly, under-sown crops should receive a top dressing of nitrogenous fertilizer at a rate of 20–50 kg/ha. These top dressings of nitrogen are especially important when the crop is to be grazed or an application of nitrogen may be indicated in the spring. However, care must be taken with spring applications not to apply too much as excessive nitrogen will encourage lodging.

Roguing

The initial roguing can be done soon after plant establishment; any identifiable incorrect species can be removed. The stage just before the start of anthesis is the ideal time to check isolation and also the standards for cultivar purity according to the method required nationally.

Harvesting

The optimum stage to harvest is when the seed borne on the older tillers shows signs of maturity and ripeness; the seed yield from these older tillers with their longer inflorescences normally produces a greater seed weight than subsequent tillers. Seed ripeness is indicated by a doughy texture, as indicated by the thumb nail test, also by their fading green pigmentation; the anthocyanin pigmentation becomes more apparent in some cultivars. The seed crop can be harvested by combining direct. However, where climatic conditions allow it is more usual to swath the crop and dry in windrows before picking up. The optimum seed moisture content at the start of swathing is 45% for tetraploids and 43% for the diploids. If the seed crop is to be combined direct, then the moisture contents should be 5% lower for each of these ploidy types.

Drying

Ideally, the seed crops which have been swathed should be picked up with a combine when their moisture content has reduced to 14%, which is the highest moisture content for safe storage. When the crop is to be combined direct the moisture contents should be down to 40% for tetraploids or 38% for the diploids. Seed lots of tetraploid cultivars can lose their viability within a day when kept above 14% moisture content, although to a lesser extent for the diploids. When seed lots are to be further dried using on-floor systems ventilation should commence immediately. For optimum seed quality the moisture content must be reduced according to Table 9.1.

Heat should not be applied initially. The drying air can be reduced to 65% humidity towards the final drying stage. The seed lot should be turned at least once daily to ensure that the upper seed layer is reduced to 20% moisture content within 5 days of the start of drying. When continuous-flow flat-bed dryers are used the seed depth should not exceed 15 cm, otherwise moisture gradients will form and there is a strong possibility of the bottom seed layer overheating. After drying, the seed lot should be allowed to cool before being transferred to the seed store. When seed lots are bulk stored their temperature should be monitored to ensure that heating does not occur.

Table 9.1. Airflow drying rates for Italian ryegrass at three given seed moisture contents with maximum temperature of the drying air.

	Moisture content of seed (%)		
	35	40	45
Air flow (m^3/min)	16.2	21.5	27.8
Max. temp of drying air (°C)	54	49	38

Seed processing

The seed lots can be upgraded by an air/ screen cleaner; some seed lots may require further upgrading either by an indent cylinder or gravity separator.

Pests and pathogens

The important seed-borne pests and pathogens of ryegrasses are listed in Table 9.2.

Table 9.2. Pests and pathogens of the ryegrasses; these pathogens may also be transmitted to the crop by other vectors.

Pest or pathogen	Common name
Claviceps purpurea	Ergot
Drechslera spp.	Leaf spots
Fusarium spp.	Fusarium
Gloeotinia granigena	Blind seed disease
Clavibacter rathayi	Galls (possibly associated with nematode galls)
Anguina spp.	Nematode galls
RGCV	Ryegrass spherical cryptic virus

Festuca pratensis Huds. (Meadow grass, Meadow fescue)

Origins and types

The *Festuca* species' main centre of variation is considered to be Western Europe, where the genus has long been associated with human habitats (Hartley and Williams, 1956).

Cultivars of *F. pratensis* are mainly classified according to the earliness or lateness of the emergence of inflorescence.

The main distinguishing characters of cultivars are:

- length of flag leaf: short, medium or long;
- width of flag leaf: narrow, medium or wide; and
- length of longest stem, including inflorescence: short, medium or long.

Seed production

The seed production methods are similar to those for cocksfoot (*Dactylis glomerata*; see below). The seed is readily shed.

Pests and pathogens

The main seed-borne pests and pathogens of *Festuca* spp. are listed in Table 9.3.

Pest or pathogen	Common name
Claviceps purpurea	Ergot
Cochliobolus sativus	Spot blotch
Epichloe typhina	Cat's tail disease
Fusarium spp.	Root rot
Ustilago spp.	Smut
Anguina agrostis	Nematode gall

Table 9.3. Pests and pathogens of *Festuca* spp.; these pathogens may also be transmitted to the crop by other vectors.

Bigeneric Hybrids within the *Lolium–Festuca* Complex

The combinations and traits of combinations between these genera have been described and discussed by Borrill (1976). There are cultivars available, some of Scandinavian origin. UPOV publishes Test Guideline 243 for *Festulolium* (see Appendix).

Alopecurus pratensis L. (Foxtail, Meadow Foxtail)

Origins and types

This species originates in Western Europe but has gradually spread across to Asia. Although a common grassland species it is not very widely cultivated for seed production. The species has naturalized in parts of North America, Australia and New Zealand. It is mainly cultivated as a pasture grass in areas of relatively high rainfall.

The main seed production areas are in central Europe and northwestern USA. Meadow foxtail seed production is very similar to Italian ryegrass, as described above. This includes the recommendations for isolation, previous cropping, site, crop establishment and seed processing.

Pollination, isolation and harvesting

Meadow foxtail is cross-fertilized. The seed crop is usually secured by combining direct although this should be carefully timed to obtain the maximum yield of high quality seed because the seed heads ripen over a long period and can shed easily. The cutter bar should be set relatively high; however, it is important that the remaining plant material is subsequently cut and removed from the site so that new tillers are produced for seed production in the following year.

Pests and pathogens

The main seed-borne pathogens of *Alopecurus pratensis* are listed in Table 9.4.

Dactylis glomerata (Cocksfoot, Orchard Grass)

Origins and types

According to Borrill (1976), cocksfoot is a good example of a polyploidy complex. The species

Table 9.4. Pathogens of *Alopecurus pratensis*; these pathogens may also be transmitted to the crop by other vectors.

Pathogen	Common name
Claviceps purpurea	Ergot
Dilophospora alopecuri	Twist
Sclerophthora macrospora	Downy mildew

has a wide range of Eurasian tetraploids with some enclaves of diploids; wild hexaploids can be found in North Africa.

For the classification of cultivars, see UPOV Test Guideline 031 (see Appendix).

Site

Seed production of cocksfoot is similar to seed production of Italian ryegrass seed. The usual practice for cocksfoot seed production is to allow for the production field to be occupied for a 4-year period; this period consists of the initial establishment year followed by 3 years of seed production.

Seed crop establishment

The crop is usually spring sown in rows from 45 to 60 cm apart at a sowing rate of 8 to 10 kg/ha; the wider row widths are adopted where water availability is known to be relatively low. It should be noted that if the seed crop is to be swathed at harvest then a closer row spacing should be used so that the cut material sits off the ground when field drying. A cover crop may be sown in the initial year but should not be a strong competitor with the cocksfoot and ideally a species with a relatively early harvest. The inter-rows should be tilled following the completion of the cover crop. The crop should be closed up for seeding at the start of the second year.

Roguing

The optimum roguing time is at the start of anthesis; this is also a useful stage to confirm that the isolation is satisfactory.

Harvesting

A decision on the most appropriate method of seed harvesting depends on the prevailing climate. In those areas with relatively dry weather conditions at harvest time it is preferable to swath the crop and pick up when the seed moisture content is down to 14%. Alternatively, the crop can be combined direct.

Drying and seed processing

These operations are as described above for Italian ryegrass.

Pests and pathogens

The main seed-borne pests and pathogens of *Dactylis* are listed in Table 9.5.

Phleum pratense (Timothy)

Origins and types

This species is a native of Europe and was introduced into North America by settlers. It is said to have gained its common name timothy after an enthusiastic farmer, Tim Hanson, who introduced it into the Southern States of the USA from New England.

Timothy cultivars are classified according to the start of anthesis: early or late.

Table 9.5. Pests and pathogens of *Dactylis*; these pathogens may also be transmitted to the crop by other vectors.

Pest or pathogen	Common name
Claviceps purpurea	Ergot
Colletotrichum graminicola	Anthracnose
Dilophospora alopecuri	Twist
Drechslera spp. (syn. *Cochliobolus sativus*), which is the sexual state of *Bipolaris sorokiniana*	Blotch, 'early senescence'
Gibberella avenacea	Mould and stem rot
Sclerophthora cryophyla	Downy mildew
Ustilago striiformis	Stripe smut
Ustilago sp.	Loose smut
Clavibacter rathayi	Yellow slime disease, Rathay's disease
Anguina agrostis	Nematode gall

The cultivars are generally identified under field crop conditions according to:

- flag leaf length: short, medium or long;
- flag leaf width: narrow, medium or wide; and
- stem length, including inflorescence: short, medium or long.

For the classification of cultivars see UPOV Test Guideline 034 (see Appendix).

Timothy is frequently incorporated in leys. The crop management for seed production is essentially the same as outlined above for cocksfoot; the notable differences are as follows.

Seed crop establishment

Timothy generally succeeds at a higher plant density per unit area than cocksfoot. Plant populations are established at an inter-row spacing of 10–20 cm; a wider spacing of 45–60 cm is recommended where weed competition is likely to be significant and difficult to control. The sowing rates are 8 kg and 6 kg/ha, respectively, for these row spacings. Some growers sow a row marker crop (such as mustard) to indicate the rows so that cultivations for weed control can start earlier. The marker crop can be eradicated either by a suitable herbicide or otherwise hoed out. Anthesis of timothy can be expected to be approximately a fortnight later than cocksfoot.

Harvesting

The seed crop maturity and time of harvesting is indicated by the seed changing to a brownish colour; this may also coincide with seed shattering from the inflorescence tip. When harvesting by combine it is prudent to set the combine with a wide clearance and only gather the most mature and ripe seeds. A second harvest is made later to gather the bulk of the seed crop.

Pests and pathogens

The main seed-borne pests and pathogens of timothy are listed in Table 9.6.

Table 9.6. Pests and pathogens of timothy; these pathogens may also be transmitted to the crop by other vectors.

Pest or pathogen	Common name
Cladosporium phlei	Timothy leaf spot
Claviceps purpurea	Ergot
Drechslera phlei	Leaf streak
Drechslera spp.	Leaf streak diseases
Ustilago striiformis	Stripe smut, timothy smut
Anguina agrostis	Ear cockle

Bromus inermis Leyss. (Soft Brome, Brome Grass, Smooth Brome)

Origins and types

There are two types of *Bromus inermis*; the North American native species and the Eurasian type, which was introduced into North America. *Bromus inermis* is a native of the Eurasian temperate zone forests. The evolution of *Bromus* in North America has been outlined by Borrill (1976). The species is drought tolerant and is also winter hardy.

The Southern USA group has more vigorous underground root-stocks than the Northern group and it produces earlier spring growth. There has been some development of European cultivars:

- *Bromus arvensis* L. (field brome);
- *Bromus carinatus* Hook et Arn. (California brome);
- *Bromus sitchensis* Bong. (Alaska grass);
- *Bromus wildenowii* Kunth. (rescue grass, prairie grass); and
- *Bromus auleticus* Trin. Ex Nees – a relatively new selection of interest for grazing in southern Brazil. The *Bromus* species are largely used for the establishment of leys.

For the classification of cultivars see UPOV Test Guideline 180 (see Appendix).

Seed crop production

Bromus seed is produced as described for cocksfoot.

The main seed-borne pests and pathogens of *Bromus* spp. are listed in Table 9.7.

Arrhenatherum eliatus (L.) Beauv. Ex J.S. et K.B. Presl. (Tall Oat Grass, French Ryegrass, False Oat Grass)

Origins and types

This grass species is a native of Europe, West Asia and North Africa. It subsequently spread to central and eastern Europe where it is widely cultivated.

Seed production

The seed production of tall oat grass is similar to the production of *Dactylis glomerata*. The seed yield of this species is improved when the crop rows are up to 1 m apart.

Harvesting

Seed of tall oat grass sheds very readily. It should be harvested by combine when the inflorescence straw turns yellow and before seed shedding starts at the top of the panicle. This stage can be determined by striking several panicles with the palm of the hand; if the inflorescences yield seed then the bulk of seed can be harvested by combine. The seed should not be threshed too vigorously otherwise the naked caryopsis may result in the reduction of potential germination.

Table 9.7. Pests and pathogens of *Bromus* spp.; these pathogens may also be transmitted to the crop by other vectors.

Pest or pathogen	Common name
Claviceps purpurea	Ergot
Cochliobolus sativus	Spot blotch, root rot, seedling blight
Drechslera gigantea	Zonate eye spot
Gloeotinia granigena	Blind seed spot
Pseudoseptoria bromigena	Halo spot
Pyrenophora bromi	Brown leaf spot
Pyrenophora semeniperda	Sterility disease
Tilletia fusca	Kernel smut
Tilletia tritici	Bunt, rough-spored bunt
Ustilago bullata	Head smut
Ustilago striiformis	Stripe smut
Anguina spp.	Nematode galls

Pests and pathogens

The main seed-borne pests and pathogens of tall oat grass are listed in Table 9.8.

The Prairie Grasses

This group of grass species is of particular interest in areas which have a climate consisting of very hot summers followed by severe winters. Organized seed crop production of this group is a relatively new technology compared with earlier methods of harvesting seed from long-established pastures. Some species have received attention as ornamental grasses while some have been cited as possible sources of biofuels.

Agropyron spp. (The Wheat Grasses, Quackgrass)

Origins and types

There are ten *Agropyron* spp. of economic importance as prairie grasses:

- *A. desertorum* (Fisch. Ex LK.) Schult. (standard crested wheat grass);
- *A. cristatum* (L.) Gaertn. (fairway crested wheat grass);

Table 9.8. Pests and pathogens of tall oat grass; these pathogens may also be transmitted to the crop by other vectors.

Pest or pathogen	Common name
Tilletia controversa	Dwarf bunt
Uromyces arrhenatheri	Stem smut
Ustilago avenae	Loose smut
Anguina spp.	Nematode galls

- *A. dasystachum* (Hooker) Scribn. (northern wheat grass);
- *A. elongatum* (Host) Beauv. (tall wheat grass);
- *A. inerme* (Scribn et J.G. Smith) Rydb. (beardless wheat grass);
- *A. intermedium* (Host) Beauv. Ex Banmg. (intermediate wheat grass);
- *A. riparium* Scribn. et J.G.Smith (stream bank wheat grass);
- *A. smithii* Rybd. (western wheat grass);
- *A. trichophorum* (LK) K. Richter (pubescent wheat grass); and
- *A. trachycaulum* (LK) Malte ex A.F. Lewis (slender wheat grass).

The seed production of *Agropyron* spp. is collectively described below; any differences in requirements of an individual species are pointed out.

Site

There should be a minimum period of 2 years between any wheat grass species. Special attention should be given to ensure that weed grass species, especially *Agropyron repens*, have been eradicated and also that the site is free of dormant seeds of *Avena fatua* (wild oat).

Pollination and isolation

All of the above *Agropyron* spp. are cross-fertilized except for *A. trachycaulum*, which is self-fertilized. The isolation distances for cross-fertilized species are 200 m for fields of 2 ha or less, otherwise 100 m for larger fields if the seed to be produced is to be further multiplied. *Agropyron trachycaulum* requires a 3 m-wide minimum isolation strip.

Seed crop establishment

Low plant densities are generally recommended for wheat grass seed production. Inter-row distances of up to 60 cm are generally adopted, with even wider inter-row distances where there are likely to be water shortages during the growing seasons. The general sowing rates for the wheat grasses are generally 2–3 kg/ha.

The seed production fields should have the remaining straw, thatch and plant debris removed by a range of operations including grazing, mowing and raking at the end of the seed production season. The soil crust in inter-row areas should be lightly broken up by rotary cultivators in the spring, followed by an application of a nitrogenous fertilizer at a rate of approximately 100 kg/ha; this top dressing becomes more essential as the site enters its third and fourth seed production years.

Roguing

This activity should include the regular removal of invasive grasses and wild oat plants that may occur in the seed crop. The fields should be rogued before the start of anthesis; isolation should also be checked at this time.

Harvesting

The seed crop can either be swathed and dried in windrows or combined direct. The optimum seed stage for swathing is at the soft to hard dough stage. The ideal stage for combining the seed crop is determined when the seeds have just become hard; however, harvesting should then take place immediately as the crops shatter very readily. The combine should be set to cut as high as possible, leaving the straw to be cut later as a separate operation.

Drying

Seed picked up from the swath usually has a moisture content of 14% or less, which is a safe moisture level for open storage. Seed lots harvested by combine should have green material removed immediately following the harvest and be subsequently dried with frequent turning until it has reduced to 14% moisture content.

Seed processing

Some of the wheat grasses' seeds require de-awning, otherwise they can be cleaned with an air/screen cleaner.

Pests and pathogens

The main seed-borne pests and pathogens of *Agropyron* spp. are listed in Table 9.9.

Andropogon spp. (The Bluestem Grasses)

Origins and types

This group of grasses has developed as the main indigenous species of the North American prairies. The three bluestem grass species considered as important forage crops are:

- *A. geradii* Vitm. (big bluestem, turkey foot bluestem) – it is the official state grass of Illinois, USA, and is generally considered to be a very good forage plant under prairie conditions;
- *A. hallii* Hach. (sand bluestem); and
- *A. scoparius* Michx. (little bluestem).

Table 9.9. Pests and pathogens of *Agropyron* spp.; these pathogens may also be transmitted to the crop by other vectors.

Pest or pathogen	Common name
Claviceps purpurea	Ergot
Puccinia graminis	Black rust
Tilletia controversa	Dwarf bunt
Tilletia tritici	Bunt, rough spored bunt
Ustilago bullata	Head smut, loose smut
Ustilago tritici	Loose smut
Ustilago turkomanica	Head smut
Corynebacterium agropyri	Yellow gum disease
Anguina spp.	Seed galls

Some of the bluestem grasses are cultivated as ornamentals.

Seed production

Each of the above three bluestem species is cross-fertilized. The seed crop production is similar to that of *Andropogon*. The seed production site should be adequately supplied with irrigation water, which is especially important during anthesis otherwise there may be an excessive quantity of infertile spikelets. Seeds develop relatively late in the production season. The harvested seed should be passed through a de-awner to assist the breaking up of fertile and infertile spikelets prior to further processing. Following de-awning the seed lot can be upgraded by an air/screen cleaner.

Pests and pathogens

The main seed-borne pathogens are *Sphacelotheca* spp. (kernel smuts); these may also be distributed by other vectors.

Phalaris spp. (Canary Grasses)

- *Phalaris arundinacea* L. (reed canary grass); and
- *Phalaris aquatica* L. (harding grass, phalaris).

Types

Some cultivars have been developed that have a reduced shedding habit.

Seed production

The seed production procedures are similar to the requirements for *Agropyron*.

Pollination and isolation

Both of these species are cross-fertilized.

Harvesting

The seed-bearing spikes tend to mature and ripen from the top downwards; the optimum stage for harvesting is when approximately half of the seeds on the average spike have turned brown.

Pests and pathogens

The main pathogens and a nematode which are seed transmitted include:

- *Claviceps* spp.: ergot;
- *Ustilago longissima*: brown stripe smut; and
- *Anguina agrostis:* grass seed nematode.

These may also be distributed by other vectors.

Elymus spp. (Wild Rye)

The two species of economic and food security value are:

- *Elymus junceus* Fisch. (Russian wild rye); and
- *Elymus canadensis* L. (Canadian wild rye).

The seed production requirements for these two *Elymus* spp. are similar as described above for *Agropyron* spp. Both species are cross-fertilized.

Harvesting

Russian wild rye seed harvesting operations should commence relatively early because the maturing seed of this species sheds easily; harvesting should therefore commence while the seed is in the dough stage if the crop is to be swathed, otherwise when the seeds begin to harden if the seed crop is to be combined direct.

Canadian wild rye seed does not shed as readily as Russian wild rye and there is a wider window of opportunity for the successful harvesting of the seed crop. The seed of this species has awns and therefore should be passed through a de-awner prior to further upgrading with an air/screen machine.

Tropical Grasses

The use of purposely developed and maintained pastures for livestock grazing and supply of stored fodder such as hay and silage has become an important development in recent years in the tropics and subtropics. This is a result of more sustainable systems of livestock for the production of milk and its products as well as meat. Other reasons for the development of managed tropical pastures are the further development of leather and hide industries, production of organic manure and the continued employment of some livestock for traction purposes.

The cytotaxonomic backgrounds and history of several tropical and subtropical grasses which mainly originate in Africa have been discussed by Harlan (1984b). Interestingly, most of the species are now also used for livestock in South America.

The following tropical grasses are dealt with in this section.

- *Chloris gayana* Kunth. (Rhodes grass);
- *Setaria sphacelata* (Schumach.) Staph. et C.E.Hubb. (South African pigeon grass, golden timothy grass, setaria);
- *Sorghum sudanense* (Piper) Staph. (Sudan grass);
- *Sorghum bicolor* × *sudanense* (Sudan grass);
- *Panacetum milaceum* L. (common millet, proso millet);
- *Brachiaria* spp.;
- *Brachiaria decumbens* Staph. (signal grass, Surinam grass);
- *Brachiaria humidicola* (Randle) Schweichart (Korinovia grass);
- *Brachiaria ruziziensis* Germaine et C. Everardt (signal grass, ruzigrass); and
- *Brachiaria mutica* (Forsk.) Staph. (para grass).

Chloris gayana Kunth. (Rhodes Grass)

Origins and types

Rhodes grass is a perennial species although there are some annual ecotypes. This species originates from East Africa and southern Africa and was introduced into cultivation by Cecil Rhodes. It is cultivated for grazing, silage and hay and has become an important crop in parts of Asia, Australia and Central and South America. There are both diploid and tetraploid cultivars. Some of the more recently developed cultivars have resistance to root knot eelworm.

Site

Rhodes grass has a relatively high tolerance of saline soils and also has good drought resistance. It succeeds in a wide range of soils although not in heavy clayey soils. This species will succeed in soils with a wide pH range, therefore the need for liming during site preparation is very unlikely. The soil fertility levels should be checked during site preparation; if either phosphorus or potassium levels are low then these should be corrected as also should be the available calcium and sulfur. Micronutrients can be deficient on some tropical soils, therefore the levels of boron, copper and manganese should be determined and applied when found necessary.

There should be a minimum of 2 years between successive crops of Rhodes grass, unless the same cultivar and same or lower seed class is to be grown. Care must also be taken to eliminate any carry-over or weed grasses whose seeds have similar morphological characters and/or size to *C. gayana*. Appropriate cultural preparations should be made to eradicate all other grass and noxious weed species from the site during the preparations, including weed species that compete especially during the Rhodes grass establishment.

Pollination and isolation

Rhodes grass is cross-pollinated; for fields of 2 ha or less, isolation should be 200 m or 100 m depending on whether the seed crop is intended for further multiplication. For fields over 2 ha the distances should be 100 or 50 m, respectively. The diploid and tetraploid cultivars do not inter-cross but some infertility is likely when anthesis coincides between neighbouring seed crops.

Seed crop establishment

According to Purseglove (1974), germination of this species will improve if the seed is stored in satisfactory conditions for 1 year prior to sowing. High plant densities result in a more even seed maturity and ripening (Boonman, 1973). The lower sowing rate of 1 kg/ha of pure, high germinating seed can be sufficient at a row spacing of 100 cm, although higher seed rates at 25 cm row spacings are preferred by some seed producers.

Roguing

Unless the seed crop is produced at a low plant population the removal of off-type plants is not very practical except in the crop's early establishment.

Harvesting

The system of harvesting the seed crop depends on the available labour and also the prevailing weather. The earlier maturing and ripened seed will generally yield seeds with the highest potential viability. Therefore where hand harvesting of individual seed heads is practised these earlier seeds should not be lost. Hand picking at weekly intervals will maintain the harvesting momentum. The cut seed heads may be further dried on a drying floor, otherwise the seed heads can be cut with the straw and stooked *in situ* for further drying.

Drying

Hand-picked seed will generally have a moisture content of approximately 14%. This is the maximum safe moisture content for open storage. All seed, regardless of harvest method, should be dried down to this level, with the seed thinly spread on a ventilated floor and turned regularly.

Processing

The harvested seed material may be hand-threshed or swathed and picked up by combine; otherwise the crop can be combined direct or harvested by stripper to beat the seed heads and separate the seeds. The seed lot contains a range of materials including parts of the original inflorescences. Therefore the seed cleaning process requires extra attention to the settings of seed-cleaning machinery. A well-cleaned seed lot can be expected to contain approximately 30% fertile seed of which 70–80% is viable.

Setaria sphacelata (Schumach.) Staph. et C.E.Hubb. syn. *S. anceps* Staph. Ex Massey (South African Pigeon Grass, Golden Timothy Grass, Setaria)

Origins and types

This species is a major component of the natural tropical grasses of southern Africa. There are two morphological types: an annual type and also a tillering form with rhizomes. Ecotypes have been selected and further improved for cultivated leys and pastures. There are diploid and tetraploid cultivars.

Seed production

This species' cultural requirements are similar to those of *Chloris gayana*. The main difference is that the optimum inter-row spacing is 30–50 cm (Boonman, 1973). Setaria has a longer anthesis period and care should be taken when estimating harvest times. There are no attached awns, therefore seed processing is more straightforward.

Pathogens

The important seed-borne pathogens of setaria are *Tilletia* spp., the bunts; these may be also be spread by other vectors.

The Sudan Grasses and Fodder Millet

- *Sorghum sudanense* (Piper) Staph. (Sudan grass);
- *Sorghum bicolor* × *sudanense* (Sudan grass); and
- *Panacetum milaceum* L. (proso millet).

Origins and types

The Sudan grasses originate from the Sudan but have undergone significant development, with several available cultivars. The improved materials are important fodder crops in southern Africa and continue to be further developed in the USA for fodder production in drier and hotter environments. Some of the plant breeding work has also been directed towards production of cultivars with low dhurrin content; this chemical breaks down to prussic acid and is a character of the early growth of crops used for grazing. Other breeding work has concentrated on improving the palatability and digestibility of these grasses by using the 'brown midrib' character from other grass types. All of these species are annuals.

Seed production

The three species are dealt with together here.

Site

The Sudan grasses tolerate both high temperatures and drought. They succeed on soils with a pH of approximately 6.5. Johnson grass, *Sorghum halepense*, is a difficult weed of these three species. Although it is sometimes encouraged to become established in some areas for fodder, it rapidly becomes a serious weed of the cultivated grasses. The seeds of Johnson grass are almost impossible to separate from the crop species when processing.

Pollination and isolation

The *Sorghum* species are cross-fertilized, whereas the millet is partially cross-fertilized. All three species therefore require adequate isolation. Isolation distances are usually 200 m for plots of 2 ha and less, with 100 m for larger areas. These specified isolation distances are usually halved when intended production is for fodder crop seed.

Seed crop establishment

The seed sowing rates are 4 kg/ha in rows 1m apart for Sudan grass and 14 kg/ha in rows 20 cm apart for proso millet.

Roguing

Some roguing can be done soon after seed crop establishment, otherwise harvested seed quality is very much dependent on a clean seedbed and high quality stock seed.

Harvesting

The seed harvests of these three crops should be concentrated on securing the maximum yields from the main tillers. The method of harvest will not only depend on the scale of production but on the prevailing weather conditions. Under ideal conditions two cuts per crop of sorghum can be achieved. Swathing is generally recommended in areas with relatively rain-free post-cutting conditions. Where the weather is less reliable it is usually preferable to harvest direct with a combine.

Brachiaria spp.

- *Brachiaria decumbens* Staph. (signal grass, Surinam grass);
- *Brachiaria humidicola* (Randle) Schweichart (Korinovia grass);
- *Brachiaria ruziziensis* Germaine et C. Everardt (signal grass, ruzigrass); and
- *Brachiaria mutica* (Forsk.) Staph. (para grass).

Origins and types

Most of the above *Brachiaria* spp. originated from Africa where they have been widely adopted for grassland establishment; they have also become important in tropical South America as grazing and fodder crops, especially in the milk-producing areas.

Brachiaria mutica is a native of poorly drained sites in both Africa and South America and is still used in these conditions; *B. decumbens*, which originates in tropical Africa, succeeds best in areas of high humidity.

Site

The soil's phosphorus and potassium status should be determined and brought up to satisfactory levels where indicated. The crop responds well to nitrogen and amounts of up to 150 kg/ha can be advantageous. In areas of relatively high rainfall this can be applied in two halves at 3- week intervals.

Isolation

These *Brachiaria* species are apomictic, i.e. they reproduce from an unfertilized egg or associated somatic cells. Therefore the only isolation required is a gap of 3 m or more to avoid admixture at planting and harvesting.

Seed crop establishment

Brachiaria mutica is a very shy seed producer and as an alternative can be propagated vegetatively.

The seed crops are therefore established either by sowing seed in rows 30 cm apart or by planting propagules at 30 cm apart in rows of the same distance apart. The propagules can either be stem cuttings or small divisions of mother plants.

Roguing

Once established the plants should be checked for obvious off-types or admixture.

Harvesting

The crop is harvested when the seed has become difficult to mark with the thumb nail; there are few other signs that the seed crop can be harvested. The seed can be harvested by combining direct; any lodged panicles can be gathered by setting the combine table low. Suction harvesters can be used following the combining operation to secure the maximum harvest.

Drying

Brachiaria decumbens seed should be gradually dried down to less than 9% moisture content. As the germination percentage increases after approximately 1 year of storage it is important that the optimum moisture content be achieved at the start of this period to ensure satisfactory overall seed quality.

Seed processing

Brachiaria seed can be cleaned with an air/screen cleaner.

Further Reading

Grasses

Bogdan, A.V. (1979) *Tropical Pasture and Fodder Plants.* Longman, New York, USA.

FAO Grassland Index. Available at: www.fao.org/ag/AGP/AGPC/doc/pasture/forage.htm (Accessed 20 March 2011).

Humphreys, L.R. and Partridge, I.J. (1995) *A Guide to Better Pastures for the Tropics and Sub-tropics.* Tocal College, Patterson, NSW, Australia.

McDowell, R.W. (ed.) (2008) *Environmental Impacts of Pasture-based Farming.* CAB International, Wallingford, UK.

Suttie, J.M., Reynolds, S.G. and Batello, C. (eds) (2005) *Grasslands of the World.* FAO, Rome.

10

Leguminosae – Temperate and Tropical Forage Legumes

Leguminosae – Temperate Forage Legumes

The Clovers

Clover species suitable for temperate regions are:

- *Trifolium repens* L. – white clover;
- *Trifolium pratense* L. – red clover;
- *Trifolium hybridum* L. – alsike;
- *Trifolium fragiferum* L. – strawberry clover;
- *Trifolium alexandrinum* L. – berseem, Egyptian clover;
- *Trifolium incarnatum* L. – crimson clover;
- *Trifolium resupinatum* L. – Persian clover;
- *Trifolium semipilosum* Fres. – Kenya clover, Kenya white clover;
- *Trifolium vesiculosum* Savi. – arrow-leaf clover; and
- *Trifolium subterraneum* L. – subterranean clover, sub clover.

Origins and types

The centre of diversity for the clovers is the eastern Mediterranean. Subsequent settlements in Australasia and North America have resulted in a wider distribution and further cultivar development in new environments. The history and development of the clovers has been described in detail by Pasumarty *et al.* (1995). It is interesting to note that their natural ability to increase the available soil nitrogen has only recently become important with the increased cost of nitrogenous fertilizers followed by trends to more sustainable grazing and fodder systems for livestock. This has also been demonstrated in mixtures for leys and also in the 'grassing down' of orchards.

Trifolium repens L. (White Clover)

Cultivars and types

The cultivar types for this species include:

- the relatively short-lived and large-leaved group; these are also referred to as 'Dutch white' or Ladino white; and
- the longer lived and small-leaved group; generally known as 'wild white clover'.

Cultivar description

The picric acid laboratory test is used to determine the proportion of plants which have cyanogenic glucosides in their leaves, described by Dayday (1955).

The classification of white clover is detailed in UPOV Test Guideline Number 038 and the classification of subterranean clover is detailed in UPOV Test Guideline 170 (see Appendix).

Seed production

Choice of site

Clover seed can survive in the soil over a long period, therefore a minimum break of 4 years is recommended since a previous white clover crop and sowing a seed crop, although this break period can be shortened if the previous crop was the same cultivar.

Important weeds include the following:

- *Cuscuta* spp. – dodder;
- *Carduus* spp. – thistles;
- *Cirsium* sp. – thistles;
- *Cerastium vulgatum* – mouse-eared chickweed;
- *Chenopodium album* – fat-hen;
- *Geranium* spp. – cranesbill;
- *Medicago lupulina* – trefoil;
- *Melandrium* sp. – campion;
- *Myosotis arvensis* – forget-me-not;
- *Plantago major* – broad-leaved plantain;
- *Prunella vulgaris* – self-heal;
- *Rumex* spp. – docks and sorrels;
- *Sherardia arvensis* – field-madder;
- *Stellaria media* – chickweed; and
- *Trifolium dubium* – suckling clover.

The necessary work and care required to commence with a weed-free seedbed are well repaid in respect of the purity of subsequent seed crops. This is especially important as these clover species can normally continue producing a commercially valid seed crop for several years; for example the smaller leaved wild white clover type can produce an economic seed crop for up to 10 years from its initial sowing. There are four basic systems for seed production of white clover, these are:

1. Taking seed from a pasture. Seed yields are likely to be low and not very pure.
2. Producing the seed from a pure stand of white clover managed for the maximum yield of pure seed.
3. Sow with a non-aggressive grass species; this provides a better pasture or forage crop when the field is not closed off during the seed production and harvesting period.
4. Seed of some local cultivars is occasionally produced from very old and long-standing pastures that have developed a wild white clover sward as a result of continual grazing. Although this results in low seed yields, the seed can be used to maintain stocks for further multiplication.

A decision on the above options will depend on the relative value of the crop as a livestock feed or as a seed crop. Pollination rate and seed set can be increased by having one hive bee colony per hectare.

Crop establishment

There are two possible sowing times, these are:

1. Sow a pure stand in the spring on a prepared site. However, it may be considered economical to sow under a cover crop; ideally, in this case a stiff-straw cereal. In these circumstances the usual nitrogen application of the cereal is reduced so that the competition from the otherwise dominant cereal is reduced.
2. Sow in late summer (but not into autumn) so that the young plants establish before the start of poor growing conditions. Cover crops are not used in conjunction with this sowing time.

Seed crops are usually sown in the spring. The seed crop is drilled in rows approximately 15 cm apart; under English conditions the optimum rate is 1.5–3 kg/ha. If a companion grass such as meadow fescue is included, the grass sowing rate is 2–3 kg/ha. The seedbed should have a fine tilth and be firm. The sowing site should be carefully graded so that machinery can work efficiently on an even soil surface.

White clover succeeds on heavy soils with regular water supply through the growing season, although dry conditions are essential during the seed harvest. The seed can normally be harvested approximately five weeks following closing the field for seed production. The optimum environmental conditions for white clover seed production have been described by Pasumarty *et al.* (1995). Their findings could well be applied when looking for new white clover seed production areas.

The soil for white clover should have adequate available calcium; if not, a dressing of lime should be incorporated during site preparation. Available phosphorus and potassium should also be at satisfactory levels. Nitrogen may be required for crop establishment, depending on the soil's nitrogen status; however, once established there is a symbiotic availability of nitrogen from *Rhizobium* which will be sufficient. Any unnecessary application of nitrogen will restrict the clover population in the sward.

Isolation

T. repens is cross-fertilized, pollination is mainly by insects, predominantly honeybees. The recommended isolation distances are:

- Fields of 2 ha or less, 200 m for stock seed production; 100 m for lower seed categories.
- Fields greater than 2 ha, 100 m for stock seed production; 50 m for lower seed categories.

Seed crop management

The control of weeds is important; this should be done with appropriate herbicides in the absence of a suitable IPM system for weed control.

Grazing by sheep is generally considered to be the optimum system to encourage the spread of clover into any gaps, although excessive grazing will cause undesirable defoliation. In the absence of suitable livestock for grazing, the field should be trimmed followed by immediate removal of the cut material. It is essential that the stolons which produce the inflorescences are not shaded by clover or companion grass foliage.

Roguing

The variations in microclimate can influence leaflet size, therefore some allowance should be made for this when making comparisons with the cultivar's description.

The most economically and technically important weeds should be removed from the crop before they flower.

Harvesting

Anthesis lasts for several weeks, but the highest quality seed is produced from the earlier flowers in this period. The seed heads become brown as they ripen and the seed changes to a light yellow. The optimum stage for seed harvesting is when 80% of the seed heads have reached this stage. The seed is likely to germinate in the seed head in wet conditions.

The crop is cut and picked up later as combining direct is not usually feasible. The seed is threshed out of the seed pods or 'hulls' with a clover huller. These machines have a double threshing mechanism with a hulling cylinder. The hulling cylinder operates at a rate approximately 15% faster than the threshing cylinder. There are hulling attachments for stationary threshers.

Seed processing

It is essential that the threshed material is processed immediately as the mixture of seed and other plant material can heat up very quickly, even in a few hours. The material can be pre-cleaned by aspiration to separate the green material from the seed; this initial operation reduces subsequent drying time.

Seed drying

When drying by on-floor systems the air temperature should not exceed 10°C above ambient and not exceeding 35°C. The air temperature of continuous-flow driers should not exceed 38°C although for high initial moisture content the temperature should be reduced to 35°C.

Storage

White clover seed can be stored for approximately 1 year at 12% moisture content. A moisture content of 9% is recommended for longer term storage, in which case the seed lot should be at the lower moisture content from the start of its storage period.

Trifolium pratense L. (Red Clover)

Cultivars and types

Modern plant breeding has resulted in a clearer classification of cultivars into four types based on flowering time and ploidy:

- spring growth and flowering: early or late; and
- ploidy: diploid or tetraploid.

Cultivar description

- stem length at flowering time: very short, short, medium, long, very long;
- central leaflet length; upper leaf below terminal flower: short, medium or long;
- central leaflet width: narrow, medium or broad; and
- time of flowering: very early, early, mid, late or very late.

For the classification of cultivars, see UPOV Test Guideline 005 (see Appendix).

Isolation

The isolation requirements are as described above for *T. repens* (white clover). Red clover is cross-pollinated, mainly by hive and bumble bees. Diploid cultivars do not set seed when pollinated by tetraploids, or the reciprocal of these two types. The two types should have an isolation distance of at least 50 m to avoid a reduction of potential seed yield.

Choice of site

Red clover seed remains viable when left in the soil, therefore a rest period of approximately 6 years is recommended. Two important seed- and soil-borne diseases of red clover can also persist in the soil for several years; therefore a 6-year break can also assist in reducing the possibility of their infecting a new crop. However, in the absence of both diseases and when the seed is only being produced for forage crops the interval may be reduced to 2 years.

The following weeds are regarded as potentially difficult in red clover seed crops:

- *Cuscuta* spp. – dodder;
- *Carduus* spp. – thistles;
- *Cirsium* sp. – thistles;
- *Geranium* spp. – cranesbill;
- *Medicago lupulina* – trefoil;
- *Melandrium* sp. – campion;
- *Plantago lanceolata* – ribgrass; and
- *Rumex* spp. – docks and sorrels.

Trifolium pratense thrives best on a well-drained and fertile loam. The species will not tolerate waterlogged conditions although regular rainfall or supplementary irrigation is necessary. Red clover can tolerate acid soil conditions between pH 5.8 and 6.2 but in lower pH conditions a liming material should be incorporated during site preparation.

A fine but firm seedbed should be prepared; rolling prior to sowing is useful unless rain or irrigation have already firmed the soil. The early-flowering cultivars are normally sown as a pure stand. The late-flowering cultivars also generally produce satisfactory results without a companion crop, but some seed producers add a less vigorous grass; timothy (*Phleum pratense*) is used in the UK, otherwise another less vigorous grass species can be used that produces seed at the same time as the clover.

Red clover is generally sown in spring. The seed can either be drilled in rows 15 cm apart or broadcast. The sowing rates are:

- diploid cultivars 5 kg/ha; and
- tetraploid cultivars 10 kg/ha.

Wider row spacing may be used in drier areas, although lodging must be avoided so as to ensure that the seed crop does not fall into the inter-row areas making it difficult to harvest. If this wider spacing is adopted the seed rate should be reduced to 1 kg for diploids and 2 kg for tetraploid cultivars.

Roguing

Some early flowering plants can be removed from late-flowering cultivars, otherwise roguing is generally not regarded as practical for this crop. Some of the more objectionable weed species can be hand pulled while checking the crop.

Harvesting

Anthesis continues over several weeks resulting in an extended seed ripening period. Therefore the timing of the seed harvest should take into consideration as to when the maximum amount of good quality seed is obtainable. The best quality seed is generally obtainable from the earlier flowers. The weather and possibility of rains setting in should also be taken into account.

A seed head is confirmed as being ripe when it has turned brown and the seed, which can be rubbed out, is hard and dark coloured. Ideally the seed crop should be direct combined. Otherwise the crop is windrowed; when windrowing is planned, the crop is cut several days before the stage described above for combining; the seed will ripen fully while in the windrows. There can be a tendency for the seed to germinate in the head while in windrows during damp conditions. The combine should be carefully set and care must be taken to check that the seed is not being damaged during the combining process. Windrowed crops are transferred to a stationary thresher, ideally using a clover huller attachment. The small clover seeds flow very readily therefore combine casings and other equipment should be sealed to avoid seed loss.

Drying, processing and storage

The details for *T. repens* (white clover) also apply to red clover; but because red clover seeds are larger and the quantities are larger, bulk drying bins are more economic and easier to use.

The specific clover processors regarded under white clover are also used for red clover. The recommended moisture content for storing at ambient temperatures is 10%. Further work on red clover storage has been described by Wang and Hampton (1991).

Trifolium hybridum L. (Alsike)

Description and types

This clover species is a short-lived perennial. There are diploid and tetraploid cultivars.

Seed production

Crop establishment for seed production is similar to the requirements of red clover. The crop is pollinated by hive bees. The seed is usually collected from the first growth without any preceding defoliation. Harvesting is as described for red clover. Autumn grazing is beneficial otherwise all excess growth should be cut and removed prior to cessation of growth at the start of winter.

Trifolium fragiferum L. (Strawberry Clover)

This clover species is a perennial, which is cultivated for grazing in mild temperate areas including parts of Australasia and western USA. It is a perennial, with seed production generally similar to that of *T. repens* (white clover) although there are some differences in some operations. Strawberry clover is adapted to poor, dry soils and is tolerant of wet saline conditions and also alkaline soils. This species is often used in mixtures for irrigated pastures. It tolerates a soil pH range of 6.0–8.4.

Much of the development and breeding activities relate to high tolerance of sheep or cattle grazing. There are both cross-pollinating and self-fertile types; therefore the breeder of the cultivar should be consulted to confirm the required isolation and pollination requirements. Several successive seed crops can be harvested over 2 or 3 years, although the initial harvest does not occur until after the crop's first winter.

Seed crop management

This species requires the bv. *trifolii* strain of *Rhizobium*. High plant populations per unit area are more productive for seed production.

Harvesting

The calyx becomes inflated as the crop ripens and becomes a light brown to grey colour. The capsules break off from the seed head when ripe. Care should be taken to check that the seed is ready otherwise there will be too high a proportion of shrivelled or undeveloped seed in the harvested seed lot. Conversely, if delayed, a significant quantity of seed is lost by shedding. It is frequently helpful to harvest the seed crop while there is some dew still on the mother plants, which will reduce the risk of shedding.

Trifolium alexandrinum L. (Berseem, Egyptian Clover)

This clover species is an annual. Seed is produced in a similar way as described for *T. pratense*, but with the following differences. Berseem clover is more tolerant of wet conditions than red clover. It is not winter hardy but an autumn sowing can produce a seed crop in the following year; alternatively it is treated as a true annual. This species is mainly self-pollinated but requires pollinating insects to 'trip' the flower for pollination to be effective. Thus isolation is generally considered necessary.

Seed crop establishment

The seed crop is sown in rows 15–20 cm apart with a sowing rate of up to 10 kg/ha. Seed is best harvested from the initial spring growth. Autumn-sown crops may be lightly grazed, or trimmed, before the start of winter but not in the following spring.

Harvesting

The seed crop is normally direct combined. The other postharvest processing, drying and upgrading are as described for *Trifolium repens*.

Trifolium incarnatum L. (Crimson Clover)

This species is an annual but is capable of surviving some winters. The plant has an elongated inflorescence, which is either globular or ovoid.

Seed production is similar to the methods described for *T. pratense* (red clover), with the following exceptions:

- The crop is best grown on soils with medium to low fertility in order to avoid too much leaf production.
- This species has almost no seed dormancy, therefore the ripe seed will germinate in rainy conditions; however, plant breeders have produced lines which have an element of seed hardiness.

Seed crop establishment

The usual practice for seed production is to sow in the late summer, establish the crop by autumn and harvest the seed crop the following year. The seed is approximately twice the size of red clover seed, therefore sowing rates should be correspondingly higher.

Harvesting

The crop can grow to a height of 1 m and has a large proportion of non-seed material, which makes combining difficult. For this reason the seed crop is usually cut and windrowed at 75% seed hardness. The windrowed material is usually picked up and threshed by combine.

Seed processing

The threshed seed frequently requires hulling for a second time prior to further upgrading the seed lot.

Trifolium resupinatum L. (Persian Clover, Birds Eye Clover, Reversed Clover, Shaftal Clover)

Persian clover is an annual species, which is cultivated widely in Iran and the Caucasus. This species has some soil salinity tolerance; its soil pH requirements range from 5.0–8.0.

Seed production

The seed crop is sown in the autumn in some areas and overwintered. Some grazing is beneficial as it aids more uniform flowering and subsequently more uniform seed maturity and ripening. Grazing should cease approximately 1 month prior to the estimated flowering time. Pollination is mainly by honeybees or leaf-cutter bees.

Harvesting

The seed is harvested when the majority of capsules have become a light brown colour. The capsules swell when ripe and can be easily dispersed and lost as a result of wind blow. It is fairly common practice to cut and windrow the crop and pick up later.

Trifolium semipilosum Fres. (Kenya Clover, Kenya White Clover)

Origin and types

This clover species and its subspecies have developed from clovers that reached the Arabian peninsula, Ethiopia, Kenya, Malawi, Tanzania and Uganda. There are a few named cultivars that have been developed by plant breeders in Australia and Kenya. The species is a perennial, similar to white clover. Kenya clover tends to root at the nodes. It is generally adapted to soils with a pH range of 5.4–7.0 in East Africa. Interestingly, it survives best where it is heavily grazed during the summer because in mixed pastures its stolons tend to compete with the companion grasses by growing more vertically and thereby fail to make soil contact. It produces better pastures with the sward-forming grasses.

The breeding systems and relevant interspecific relationships of some of the African clover species have been discussed by Pritchard and Mannetje (1967).

Seed production

This species should have *Rhizobium* inoculation treatment when sown on most soils. It is reputed to be very specific and requires either CB 782 (Australia) or CC 2408 (Malawi) strain.

Kenya clover is cross-pollinated mainly by hive bees. In the subtropics the crop is either mown or grazed until April, or in the upland tropics until around July. The pods do not shatter as easily as most other *Trifolium* species. The seed is harvested as described for other clovers; the harvested seed should be dried as soon as possible prior to processing or any further upgrading.

Trifolium vesiculosum Savi. (Arrow-Leaf Clover)

There are several named cultivars that have been produced by plant breeders in Australia.

Recent research work by Chilean agronomists has reported that this species has special value for the areas in South America that have a Mediterranean climate, such as the Andean foothills in the central-southern region of Chile (Ovalle *et al.*, 2010). They found that the species is also very suitable for traditional rotation systems in conjunction with either a cereal or rapeseed crop.

Arrow-leaf clover is an annual, and is sown in the autumn for seed harvest the following summer. This species is cross-pollinated by bees.

Trifolium subterraneum L. (Subterranean Clover, Sub Clover)

Origins and types

Subterranean clover originated in Europe but has become an important fodder crop in Australia as well as in parts of Central and South America. It has become particularly useful because of its suitability for drier growing conditions. The species has become an important forage crop in parts of Australia where many cultivars have been produced by plant breeders.

For the classification of cultivars, see UPOV Test Guideline 170 (see Appendix).

Site

This crop occurs as an escape species or remnant from earlier forage sites. Seed can remain dormant in the soil for several years, therefore a break of at least 6 years is recommended between seed crops. Ideally a new site should be prepared for the multiplication of new or improved cultivars. The weed *Juncus bufonius* L. var. *bufonius*, commonly known as 'toad rush', is a difficult weed in some areas where this clover is grown because of its competition and also for impeding seed harvesting and processing. Research into possible methods of its control have been reported and discussed by Seidel (2010).

Pollination and isolation

Subterranean clover is self-fertilized, therefore there is only the need for minimal isolation to avoid admixture during sowing and seed harvesting.

Seed crop establishment

This clover species produces its seed in burrs, which are just above soil level. It is therefore preferable to sow the stock seed in a seedbed that has a carefully graded surface. A high plant density usually ensures a good potential seed yield because there is the minimum of exposed soil for the ripe seed or burrs to enter the soil. Sowing rates are 6–10 kg/ha, the higher rate being used for very high plant populations. Companion sowing with a grass species improves the site's grazing value but it is generally considered that grass will reduce the clover's seed yield.

Roguing

Most of the cultivars released by breeders have linked foliage markings, which are very useful when roguing the crops. The first roguing should be made in the spring and a second check made at the start of anthesis.

Harvesting

The seed is harvested once the foliage has become dry, which is an indication that the seed has hardened. Ideally a suction harvester is the best method to pick up the seed. The graded soil surface minimizes the amount of seed entering the soil.

Seed processing and drying

The seed crop is normally passed through a pre-cleaner to remove any excess debris. Further seed processing is as described for other clover species.

Pests and pathogens

The pests and pathogens of *Trifolium* spp. are listed in Table 10.1.

Medicago sativa L. and *Medicago* × var. Martyn. (Lucerne, Alfalfa)

Origins and types

The cytotaxonomic and early history of this important crop has been described by Lesins (1984). Of the two widely used common names for this species, lucerne originates from Europe (except Iberia), while the name alfalfa is in common use in North and South America as well as Iberia. It is thought that the mountain-protected fertile valleys in the Zagros and Elburz mountains and nearby irrigated oases facing the Iranian plateau are where lucerne was first cultivated.

For the classification of cultivars, see UPOV Test Guideline 006 (see Appendix).

Site

The seed crop is most successful in a site situated in an area of low relative humidity with an adequate supply of irrigation water or frequent rainfall which is little and light. The species is deep rooting and thrives best on a deep and well-drained fertile soil. A temperature above 20°C is best during anthesis and seed ripening. According to Hacquet (1990), temperatures of 20–30°C increased fertility, pollinator activity and seed set, while either drought or excessive water reduced seed yield. Ideally, the lucerne crop requires adequate soil levels of available calcium, phosphorus and potassium. The most objectionable weeds are:

- *Cuscuta* spp. – dodder;
- *Amaranthus* spp. – pigweed;
- *Brassica* spp. – charlock, mustard and rape;
- *Cenchus* spp. – sandbur;
- *Convolvulus arvensis* – bindweed;
- *Galium aparene* – cleavers;
- *Melilotus* spp. – sweet clover;
- *Rumex* spp. – docks and sorrels;
- *Sida* spp. – mallow; and
- *Sorghum halepense* – Johnson grass.

Table 10.1. Pests and pathogens of *Trifolium* spp.; these pathogens may also be transmitted by other vectors.

Pest or pathogen	Common name
Botrytis anthophila	Anther mould
Fusarium spp.	Fusarium
Kabatiella caulivora	Scorch, northern anthracnose
Phoma medicaginis var. *pinodella*	Spring black stem
Rhizoctonia leguminicola	Black patch disease
Sclerotinia sclerotiorum and *Sclerotinia trifoliorum*	Clover rot
BCMV	Bean yellow mosaic virus
AMV	Berseem mosaic virus
Ditylenchus dipsaci	Stem eelworm, clover sickness

Seed crop establishment

The seed should be treated with the appropriate inoculum of *Rhizobium* for the location.

It is important that sites with known dodder problems are avoided. The emerging lucerne crop does not compete well with weeds, although it does once it is established when it has developed a leaf rosette at soil level.

Lucerne is normally sown as a single stand (i.e. without a companion grass) for seed production. Seed is sown at the rate of 90 kg/ha in rows 90 cm apart with the aim of achieving a stand of plants 10 cm apart in the rows. In conditions which are less than optimum the higher sowing rate of 3 kg/ha in rows 35 cm apart is adopted.

Roguing stages

1. Seed crop establishment.
2. Before anthesis, in flower bud stage; width and length of central leaflet.
3. At start of full flowering; length of longest stem, including the flower head.

Pollination and isolation

Lucerne is almost totally cross-pollinated. The structure of the lucerne flower is such that it is pollinated by those bee species that are able to 'trip' the flowers; these include honeybees, although this insect is not the most efficient as they frequently avoid tripping the flowers when collecting nectar rather than pollen. Other pollinators are wild bees (their presence should be encouraged by the conservation of some rough grass areas in the seed crops' production areas). In warm areas the leaf-cutter bee and the alkali bee are used.

Harvesting

Seed of this species ripens over an extended period, therefore timing of the harvest has to ensure that the maximum amount of high quality seed is secured. The seed pods are small, spiral pods, which change to brown/dark brown when their seeds are mature. Harvesting the crop is done when approximately 75% of the pods are at this stage. The crop can be cut and windrowed or the standing crop is desiccated and harvested direct by combine after 2–3 days (a longer delay than this will lose seed following dehiscence). Grain combines are suitable provided that they have been sealed to prevent seed loss. The combine's setting is critical to prevent seed loss and/or seed damage.

As the crop is a perennial it is important to remove detached plant debris as soon as practicable in order to prevent carry-over of prevalent pests and pathogens.

Drying

The seed moisture content should be reduced to 12% for medium term storage but down to 8% for storage periods greater than 1 year. The drying air temperature should not exceed 38°C when initial moisture content is less than 20%; for higher moisture contents the drying air temperature should be below 30°C.

Seed processing

The usual practice is to use an air/screen cleaner followed by a gravity separator, which will remove weed seeds. In the event that there is dodder seed in the seed lot, the bulk can be passed through a roll separator; however, this is very likely to result in the loss of some good lucerne seed.

Pests and pathogens

The main seed-borne pathogens and pests of lucerne are listed in Table 10.2.

Onobrychis viciifolia Scop. (Sanfoin, Esparcetta)

Origins and types

Sanfoin is a perennial herbaceous plant native to Europe and western Asia, where it has been grown for centuries for fodder; the evolution and history of this species has been outlined by Smith (1984). However, in more recent times the species has not been able to compete with the popularity and agriculturists' general preference for the modern cultivars of lucerne and red clover.

Table 10.2. The main seed-borne pathogens and pests of lucerne; these pathogens may also be transmitted by other vectors.

Pest or pathogen	Common name
Cercospora zebina	Summer black stem, summer leaf spot
Colletotrichum trifolii	Anthracnose
Fusarium spp.	Fusarium
Phoma medicaginis	Spring black stem
Verticillium albo-atrum	Verticillium
Clavibacter michiganensis ssp. *insidiosus*	Bacterial wilt
AMV	Alfalfa mosaic virus
LALV	Australian lucerne latent virus
Ditylenchus dipsaci	Stem eelworm

Cultivar description

Martiniello (1992) found that height and smoothness of shoot, standard petal profile and seed colour are the most effective characteristics for cultivar descriptions.

Site

Sanfoin requires satisfactory drainage; it is not successful on wet soils. Soils with high available calcium are very suitable; for example in England it is largely cultivated on chalky soils. It has a good tolerance to drought conditions once it has passed the seedling stage and has become established. The proposed field for seed production should not have produced sanfoin for at least 4 years, but if the seed crop is intended for further multiplication then a minimum of 6 years is recommended.

Pollination and isolation

Sanfoin is self-incompatible but cross-pollinated by insects; bees are the main pollinating agent. It is estimated that each flower should be visited by a bee on at least three occasions. Pollination success can be significantly improved if up to ten honey bee hives per hectare are placed close to or within the crop.

The recommended isolation for fields of up to 2 ha is 200 m when the seed crop is intended for further multiplication and 100 m when the seed is being produced for forage crops. For fields over 2 ha, the distances are 100 m and 50 m, respectively.

Seed production

The stock sanfoin seed is usually undersown in a relatively light cereal nurse crop in spring. Autumn sowing is not recommended. The sanfoin crop is sown in rows 15–20 cm apart at a rate of 50 kg/ha of unhulled seed or 80 kg/ha for hulled seed. Hulled seed normally germinates better than unhulled.

The sanfoin crop should not be cut or grazed following the harvesting of the nurse cereal crop unless it has made excessive growth which should be removed prior to the onset of winter. The biennial form of sanfoin is initially cut for hay in the second year, and seed is harvested from the second growth. However, the perennial type should have seed harvested from the first growth. The perennial type generally produces the better seeds in its second and third years of harvest. These crops can often continue producing a seed crop for up to 7 or more years.

Roguing stages

1. At the vegetative stage: seedling height at 12 weeks; plant growth habit and canopy; size of secondary leaf; leaf colour and leaf hair type.
2. At start of anthesis: relative height of plants; time of flowering; flower colour.

Harvesting, drying and seed processing

These operations and processes are as described for red clover (*Trifolium pratense*).

Pests and pathogens

The pathogens of sanfoin and other *Onobrychis* spp. are listed in Table 10.3.

Table 10.3. Pathogens of sanfoin and other *Onobrychis* spp.; these pathogens may also be transmitted by other vectors.

Pathogen	Common name
Ascochyta onobrychidis	Anthracnose
Typhula trifolii	Snow rot
Clavibacter michiganensis ssp. *insidiosus*	Bacterial wilt
Pseudomonas fluorescens	Crown and root rot, damping-off

Ornithopus sativus Brot. (Seradella, French Seradella)

Origins and types

Seradella has been traditionally cultivated as a forage crop in the Iberian peninsula and some areas of North Africa. Plant breeders in France have developed several cultivars that now have a wider appeal and use, especially in Australasia. The modern cultivars are grown in the Australian wheat belt as an inclusion in the rotation, where it has been found to make a useful contribution to soil fertility and weed control (Graeme *et al.*, 2009).

Site

The French seradella types are very tolerant to sandy acidic soils in areas of low rainfall; the species does not tolerate waterlogging.

Pollination and isolation

This crop is generally considered to be self-fertilized; therefore the only isolation requirement is the provision of a suitable gap between crops so as to avoid admixtures during sowing and harvesting operations.

Seed crop establishment

Many seed stocks still have an individual seed enclosed in its original pod segment. These are sown at the rate of 20–40 kg/ha according to the seed lot's potential germination and field conditions. Inoculation with *Rhizobium* is usually necessary, although it is reputed to succeed without this if *Lupinus* spp. have been previously grown on the site.

Roguing

This is best done at the start of anthesis. Apart from removing off-types which are significantly different from the crop's description, the main off-types are based on flower colour.

Harvesting

The seed pods tend to drop off the mother plant as they ripen, therefore timing of harvest for maximum yield is critical. The crop is usually swathed and left in windrows and subsequently picked up with a combine. Workers in Australia report that harvesting direct with either a crop or clover harvester can be successful, although this may depend on the amount of plant debris in the seed crop.

Drying and processing

When the moisture content is greater than 10% the seed should be dried down to this level before being stored. The seed is usually left in its pod segments and upgraded in an air/screen cleaner. Further upgrading can be done by a gravity separator if required.

Pathogen

The main seed-borne pathogen of *Ornithopus sativus* (seradella) is *Colletotrichum trifolii* (anthracnose); this pathogen may also be transmitted by other vectors.

The Vetches

- *Vicia sativa* L. – common vetch, tare;
- *Vicia villosa* Roth. – hairy vetch; and
- *Vicia pannonica* Crantz. – Hungarian vetch.

Vicia sativa is the most widely cultivated of these three *Vicia* spp.

Types and cultivars

Martiniello (1992) lists several morphological characters for cultivar description and/or roguing; these are:

- height, colour and growth habit;
- attitude of shoot and axial length;
- leaf, surface and leaflets;
- inflorescence number, standard petal profile and shape of sepals forming calyx;
- length and shape of pod;
- smoothness and segmentation of marginal leaf; and
- seed colour, colour of hilum.

Site

A minimum break of 3 years is required for seed crops grown for final generations and 6 years for higher seed classes.

Isolation

Vicia sativa and *V. pannonica* are largely self-pollinated, therefore only 3 m isolation distance is necessary. *Vicia villosa* is cross-fertilized and fields of 2 ha or less should have an isolation distance of 200 m, 100 m is required for crops of earlier generations and certified seed; for fields larger than 2 ha, isolation distances are 100 m and 50 m, respectively. Bees are important pollinators of these crops and two supplementary hives per hectare are advantageous, especially for *V. villosa*.

Seed crop establishment

The seed is sown in rows 15–20 cm apart at the rate of 20–40 kg/ha, depending on field conditions. The initial plant growth should be left to produce the seed crop.

Roguing

In practice, roguing of vetch seed crops can be difficult as a result of the plants intertwining. The main characters are checked at the start of anthesis. Vetches are one of those crops where the genetic quality of stock seed is emphasized to ensure the trueness to type of the subsequent generations produced from it.

Harvesting

For *V. sativa* and *V. villosa* cutting, windrowing and picking up with a combine after some drying is possible but is very likely to result in significant seed loss in the field. Harvesting direct by combine as soon as the seeds in the lower pods are firm and show colour change is normally preferable. *Vicia pannonica* does not shed so readily and seed harvesting can be left until approximately 90% is ripe; however, it is more difficult to thresh, therefore care should be taken with combine settings to ensure that seed is not left in the pods.

Seed drying and processing

The combined seed usually has mother plant material mixed with it. The method of drying is as described for white clover. Vetch seed is significantly larger than clover and is therefore easier to handle.

The seed can usually be upgraded on an air/screen cleaner. The seed lot may contain vetch weevil larvae (*Bruchus branchialis*), if so the affected seed which is lighter in weight may be removed during the cleaning process, otherwise it should be fumigated by a licensed contractor.

Pathogens and pests

The main seed-borne pathogens and pests of the forage vetches are listed in Table 10.4.

Table 10.4. The main seed-borne pathogens and pests of vetches; these pathogens may also be transmitted by other vectors.

Pest or pathogen	Common name
Ascochyta pisi	Vetch leaf spot
Peronospora viciae	Downy mildew
Rhizoctonia solani	Damping-off
Bruchus branchialis	Vetch weevil
RLVMV	Red clover vein mosaic

Leguminosae – Forage Legumes for Warm Climates

The species listed in this section are generally cultivated in North America and the Mediterranean area; some are adapted to specific areas. Some of the species are used for soil conservation.

Lespedeza stipulacea Maxim. (Korean Lespedeza)

This species is an annual and is cultivated mainly in the southern USA, particularly to the east of the region. The crop responds well to poor soils with relatively low fertility. There are two other *Lespedeza* species, these are *Lespedeza striata* (Thunb.) Hook et Corn and *Lespedeza cuneata* (Dum.) G. Don, which are not discussed in this volume.

Site

There should be an interval of 4 years from the previous lespedeza crop before sowing a seed crop, unless the same cultivar is being grown again. The site should not have a history of dodder and should be free of *Sorghum halapense* (Johnson grass) and *Ambrosia* spp. (ragweed).

Pollination and isolation

Lespedeza is mainly self-fertilized. There are some flowers that remain closed but there are also flowers that are visited by bees. Seed yield is reputed to be significantly improved when hive bees are sited adjacent to the seed crop.

Seed crop establishment

The crop for seed production is sown in spring. Unhulled seed is sown at the rate of approximately 25 kg/ha. The initial growth is usually taken for seed, but if there is very vigorous first growth a forage crop can be secured prior to the seed harvest.

Roguing

The optimum stage for roguing is at the start of anthesis.

Harvesting

Lespedeza sheds its seed very readily when ripe, therefore harvesting should not be delayed. The crop is usually direct combined.

Drying

The seed is usually harvested with the moisture content at a satisfactory level as indicated as slightly firm to the thumb nail test for storage, but this will depend on the climate or local weather where it is being produced. However, when there is green plant debris in the harvested seed lot it is necessary to dry immediately. Drying time can be reduced by an initial pre-cleaning operation. The seed is normally left unhulled as the dormancy period is relatively short.

Seed processing

The seed lot is normally processed by an air/screen cleaner. Dodder seed can be removed with a velvet roll mill if found to be present in the seed lot.

Pests and pathogens

The main seed-borne pathogens of *Lespedeza* spp. are listed in Table 10.5.

Table 10.5. The main seed-borne pathogens of *Lespedeza*; these pathogens may also be transmitted by other vectors.

Pathogen	Common name
Rhizoctonia solani	Damping off
Uromyces lespedeza-procumbentis	Rust
Xanthomonas campestris pv. *lespedezae*	Bacterial wilt

Melilotus spp.

There are three species of *Melilotus* cultivated as fodder crops:

- *Melilotus alba* Medikus – white sweet clover;
- *Melilotus officinalis* (L.) Palles – yellow sweet clover; and
- *Melilotus indica* (L.) All – Indian sweet clover, senji, sour clover.

Origins and types

The *Melilotus* species originated in Eurasia and have long been established as fodder crops. The ability of seed to remain dormant in the soil for long periods, coupled with its spread via contaminated seed lots, has given this genus a bad reputation in some areas of North America.

Melilotus alba and *M. officinalis* are both cultivated in North America; *M. indica* is mainly grown in India. The white and yellow types are adapted to temperate climates; the yellow type is fairly tolerant of dry growing conditions. The common name of 'sweet clover' arose because cows which had received it in their feed produced sweet milk. Indian sweet clover is produced with irrigation in dry areas of India on soils that are neutral to alkaline.

Site

A 5-year break should be planned between a previous *Melilotus* crop when producing high category seed classes, but 2 years is usually regarded as sufficient when the seed crop is intended for production of a fodder crop. *Melilotus* spp. produce vigorous seedlings and therefore only a clean seedbed prior to sowing is normally sufficient.

Pollination and isolation

The white and yellow species are generally considered to be cross-fertilized. Their isolation requirement is 200 m for fields of 2 ha or less and 100 m for larger fields producing seed of earlier generations; for later generations these distances can be halved. Indian sweet clover is predominantly self-fertilized, therefore the isolation distances for Indian sweet clover are 50 m for earlier generations and 25 m for the later generations.

Pollination of *Melilotus* spp. is mainly by bees; supplementary hives are reported to increase seed yields significantly.

Seed crop establishment

Seed is drilled in rows from 15 to 100 cm apart; the wider spacing is used in the drier areas. The sowing rate ranges from 15 kg/ha to 4 kg/ha for the widest row spacing. Companion grasses can be included to increase the fodder value of grazed or cut herbage.

Seed crops are generally sown in the spring for harvest in the following year. The best seed crops are achieved when the crop has been grazed in the establishment year. The field is then shut off during the winter, leaving it with approximately 15 cm of uniformly trimmed growth.

Roguing

The most practical stage for roguing is at the start of anthesis when flower colour can be verified. Gaps or gangways left when sowing the crop increase the ease of crop roguing and/or inspection and can be very useful, especially for the higher seed classes as the crop can be very dense.

Harvesting

Melilotus seed readily sheds, therefore windrow harvesting is preferable to direct combining; this allows the plant material to dry further prior to threshing. The crop is ready for windrowing when approximately 60% of the pods have changed colour to brown or black; at this stage the stems will still be sappy and only a few leaves will have been shed. Chemical desiccation followed by direct combining is an option, although combining has to be completed within 3 days of desiccant application. Seed loss from shedding is significantly reduced while the crop is still damp from overnight dew. Harvesting can commence before the 60% colour change stage in more arid areas.

Drying, seed processing and storage

The material picked up by the combine from the windrows will frequently contain a high proportion of green plant debris; it is therefore advisable to pre-clean first so as to reduce the subsequent drying time. The drying details for white clover also apply to sweet clover. There is usually a proportion of unhulled seed coming from the combine, which can be reduced by scarification.

Seed cleaning and storage are as described for *Trifolium repens*.

Pests and pathogens

The main seed-borne pathogens of *Melilotus* spp. are listed in Table 10.6.

Table 10.6. The main seed-borne pathogens of *Melilotus* spp.; these pathogens may also be transmitted by other vectors.

Pathogen	Common name
Ascochyta caulicola	Stem canker, Goose-neck disease
Ascochyta lethalis	Spring black stem
Mycosphaerella lethalis	Spring black stem
Mycosphaerella davisii	Summer black stem and leaf spot
Clavibacter michiganensis	Bacterial wilt
BYMV	Bean yellow mosaic virus
LTSV	Lucerne transient streak virus
TSV	Tobacco streak virus

Lotus spp. (Trefoils)

The following three *Lotus* species are well adapted to regions with a Mediterranean summer climate. The three species are perennial. *Lotus uliginosus* is the most widely cultivated, this species producing a crown as does *Lotus tenuis*. *Lotus corniculatus* produces stems, which are able to produce adventitious roots. These three trefoils are discussed below; any exceptions for individual species are indicated.

- *Lotus corniculatus* L. – bird's foot trefoil;
- *Lotus tenuis* Waldst. et Kit ex. Willd. – slender bird's foot trefoil;
- *Lotus uliginosus* Schk. – greater bird's foot trefoil, giant trefoil.

Origins and types

These crops are believed to have been developed along the North African coast of the Mediterranean.

- *Lotus corniculatus* is more tolerant to drought but less tolerant to poor drainage.
- *Lotus tenuis* tolerates soil acidity and higher summer temperatures.
- *Lotus uliginosus* is more tolerant of wet soils.

Cultivars have been developed for more uniform seed ripening, which reduces seed loss during harvesting operations.

Site

Lotus seeds are capable of dormancy in the soil, therefore when seed is to be produced for further multiplication there should be a minimum break of 5 years, with a break of 3 years when seed production is for forage. Soils should have adequate phosphorus and potassium available, although the crop's growth will be rank and spindly in high soil nutrient regimes.

Pollination and isolation

These three trefoil species are cross-pollinated, mainly by bees. Seed yield can be significantly increased if two bee hives are allocated for a field of approximately 2 ha. Isolation for crops intended to produce seed for further multiplication should be 200 m for fields of up to 2 ha and 100 m for larger fields. These isolation distances can be halved if the crop is intended for forage seed production.

Seed crop establishment

The trefoils do not compete successfully with weeds at seedling emergence through to early plant population establishment. In addition, seeds of *Convolvulus arvensis* and *Rumex* spp. are very competitive. Admixtures of other legume fodder crop seeds, such as clovers and lucerne, can be difficult to separate when processing *Lotus* seed.

Spring sowings are preferable; seed is drilled in rows from 15 to 60 cm apart depending on the location's moisture availability. Sowing rates are 3–5 kg/ha. The advantage of the higher plant densities as a result of closer row spacing is a more uniform crop during anthesis and seed maturity.

The first seed crop is harvested the year following sowing. The crops are trimmed of excess herbage leaving a uniform stand to overwinter. The best seed is produced from the initial growth and flower production in the following year. Future seed harvests are managed in a similar way. As the crops are perennial, up to five seed crops can be achieved from one sowing.

Roguing

This operation is done at the start of anthesis to check flower colour.

Harvesting

Flowering and subsequent seed ripening of the trefoils occurs over an extended period; there is therefore a potential seed loss if care is not taken to choose the best stage for harvesting. The plants tend to remain green as the seed pods ripen. Initially the pods change colour, becoming brown to black. As the seeds harden they change to yellow-brown in *L. corniculatus*, or yellow-green in the other two *Lotus* species.

Windrowing is generally preferred to direct combining as it allows more seed to ripen within the windrow. Picking up when the seed crop is slightly damp will reduce seed loss from shedding.

Drying

The drying of the harvested *Lotus* seed is as described for white clover. The harvested seed and associated plant material will require immediate pre-cleaning to remove the maximum amount of non-seed material, although drying must not be delayed as the materials will heat up very quickly.

Seed processing

The seed processing for the *Lotus* species is the same as described for *Trifolium repens* (white clover).

Pests and pathogens

There is very little documented information regarding seed-borne pests or pathogens of *Lotus* except that the causal pathogen of crown and root rot is *Mycoleptodiacus terrestris*.

Leguminosae – Forage Legumes for the Tropics and Subtropics

There is a wide range of morphological types of plants that are cultivated as forage crops in the tropics and subtropics, including climbers and shrubs. The concept of tropical forage legume production only became established in the last decades of the 20th century and there is a lot of scope for future breeding programmes. The increased stability of farming systems and further development of livestock production in the tropics and subtropics have led to the increased activities in fodder crops produced from seed. The species adopted for this purpose are mainly 'short-day' and require 12 h or less day-length for flower initiation.

Centrosema pubescens Benth. (Centro)

Origins and type

This species originates from tropical Brazil where several other *Centrosema* species have also been found. *Centrosema pubescens* has been taken to other areas in Asia, Africa and the South Pacific and cultivated as a fodder crop; it is also used as a cover crop in some plantations. There are several available cultivars and some local landraces. This species is a vigorous climbing perennial and is capable of growing to a height of 45 cm.

Site

Sites selected for seed production should have been rested from centro for several years because up to 60% hard seeds can remain dormant in the soil; a break period of 5–6 years is usually recommended.

Pollination and isolation

Centro is self-pollinating; the only required isolation is a distance of 5 m to avoid admixture when sowing or harvesting.

Seed crop establishment

The genus *Centrosema* does not require very rich soils, but satisfactory levels of available calcium, phosphorus and potassium are beneficial.

Seed crops are best sown in rows 1 m apart at the rate of 4–8 kg/ha. The wide row spacing allows for the provision of supports for the crop, making the ripening pods more accessible; the supports can be made from locally available materials. The possibility of low germination resulting from hard seed can be overcome by either a warm water soak or light scarification of the seedcoat. The seed should be inoculated with *Rhizobium* when sowing in soils which have not previously grown centro.

Centro tends to produce some excessive vegetative growth; if this occurs either grazing or cutting will reduce it. This technique can also be used to control the time of flowering so as to adjust the start of anthesis to coincide with more suitable flowering and harvest conditions.

Roguing

The most important character when roguing is flower colour, which should be confirmed according to the landrace's or cultivar's description.

Harvesting

The seed is harvestable when the first seed pods start to open. The crops which are produced on supports are harvested in two separate pickings by hand. An unsupported seed crop is best cut and windrowed for approximately 1 week prior to picking up with a combine rather than direct combining. Windrowing reduces the amount of green material to be dealt with in the seed lot.

Drying

Despite the duration of windrowing, there will still be a large proportion of plant debris in the seed lot; therefore the drying process should commence immediately following picking up. The seed should be dried to 8–10% moisture content. Drying by spreading thinly on tarpaulins or on a concrete floor with a smooth finish in arid conditions is normally sufficient. In more humid conditions forced air drying should be used, with an air temperature not exceeding 35°C.

Seed processing

The optimum processing machine is an air/screen cleaner. The remaining non-seed material should be removed as soon as practicable in order to minimize the remaining non-seed material. The hand-harvested seed can be further cleaned by winnowing.

Macroptilium atropurpureum (DC) Urb. (Siratro, Purple Bush Bean)

This species is a climbing perennial that layers from its nodes. Although regarded as a useful tropical fodder crop it has become a problem weed in some parts of Australia. Siratro root nodules are good nitrogen producers, which assist any companion grasses.

Origins and types

The crop originates from tropical America. Several cultivars have been developed in Australia including the cv. 'Aztec', which is resistant to leaf rust.

Site

The crop requires a site with available moisture while becoming established, but develops a deep root system which is capable of sustaining the crop under dry conditions. Seed is sown at a depth of approximately 2 cm in either of these seed production systems.

Pollination and isolation

Siratro is self-pollinating, the only isolation required is a distance of 5m to minimize the possibility of admixture at sowing or harvesting.

Seed crop establishment

The higher seed yields are produced from crops grown on supports. For this system seed is sown in rows 3 m apart at a rate of 2–3 kg/ha; supports should be provided for the vines. If it is intended to harvest the seed by machine, the seed is sown at a higher plant density at shorter intra-row distances, but without supports.

Roguing

This operation should take into account the cultivar's description and any deviation from the morphological characters, which are mainly based on leaf and inflorescence. The crops produced on supports are easier to rogue efficiently.

Harvesting

The seeds shed very soon after the ripening of pods, which is indicated by their turning a dark brown colour; therefore hand harvesting the mature pods at regular intervals will secure the maximum seed yield. When machine harvesting is planned, the crop is defoliated approximately 1 month before the start of flowering. Some seed producers allow the first ripe seeds to fall into the crop canopy; the total retrievable seed yield is later combined in a single operation. The other alternative is to cut initially with the combine's cutter set high followed by a second cut at normal cutter height. Dropped seed can be either picked up by hand or with a suction harvester.

Drying and processing

The machine-harvested seed will contain a large bulk of plant debris; it will increase drying efficiency if the seed lot is processed with a pre-cleaner. Any further processing can be achieved with a winnower or an air/screen cleaner. Seed should be dried down to 8% moisture content before storage.

Pests and pathogens

The main seed-borne pathogens of siratro are listed in Table 10.7.

Table 10.7. The main seed-borne pathogens of siratro; these pathogens may also be transmitted by other vectors.

Pathogen	Common name
Diaporthe phaseolorum	Stem canker
Fusarium spp.	Fusarium rot, crown rot
Mycosphaerella holci	
Pseudomonas syringae pv. *phaseolicola*	Halo blight
BCMV	Bean common mosaic virus

Stylosanthes spp. (Stylo)

The four main species in this genus are:

- *Stylosanthes guianensis* (Aubl.) Swatz. – stylo;
- *Stylosanthes hamata* (L.) Tumb. – Caribbean stylo;
- *Stylosanthes scabra* Veg. – shrubby stylo; and
- *Stylosanthes humilis* HBK. – Townsville stylo.

Origins and types

The genus originates from tropical America from where it has been further distributed to other tropical and subtropical areas including Thailand, the Philippines and India. Research and plant breeding in Australia continues to produce new breeding lines and improved cultivars.

Stylosanthes guianensis is perennial, whereas *S. hamata*, *S. scabra* and *S. humilis* are annual.

Site

All *Stylosanthes* species are tolerant of high rainfall but require satisfactory soil drainage. The above four species are not tolerant of saline soils. Stylo seeds will remain dormant for several years in the soil; therefore there should be a break of at least 6 years before sowing a crop intended for seed production. A weed-free plot is essential to ensure satisfactory crop establishment.

Pollination and isolation

The species are generally regarded as self-pollinating although there may be some out-crossing. It is as well to consult the breeder regarding suitable isolation distances. Otherwise a clear distance of 5 m is regarded as sufficient isolation to avoid admixture during sowing or harvesting. The Indian seed regulations stipulate distances of 50 m for Foundation Seed and 25 m for Certified Seed; these Indian isolation requirements would appear to be more realistic for maintaining or even improving stylo seed quality.

Seed crop establishment

The seeds are sown in rows with the distances between the rows depending on the species. Row spacing for *S. guianensis* (stylo), which is a perennial shrub, is usually 1 m; other species are generally sown in rows 50 cm apart depending on local customs and conditions. The sowing rate is between 1 and 3 kg/ha. Some workers in Asia recommend that seed of *S. hamata* (Caribbean stylo) is broadcast in preference to drilling.

Roguing

Checking cultivar characters is best done at the start of anthesis; some of the more modern cultivars have very distinctive flower morphology and pigmentation. The breeder's cultivar description should be taken into account when roguing.

Harvesting

Seed can be collected by shaking the plants of both stylo and shrubby stylo every few days, and the seeds collected on sheets on the ground beneath the plants. Harvesting by combine is possible, but should be done when humidity is low, and this should be followed by either suction harvesting or brushing, often referred to as 'sweeping', to gather fallen seeds.

Cleaning

Seed that has been combined will usually require pre-cleaning to remove the bulk of non-seed material; the excess debris can either be removed by pre-cleaning or hand screening (i.e. sieving), using firstly a large mesh followed by a smaller mesh sieve. The final upgrading can either be done by winnowing or processing on an air/screen cleaner.

Pests and pathogens

Colletotrichum gloeosporioides (anthracnose) is a serious field and seed-borne pathogen of the stylo species.

Further Reading

American Phytopathological Society (1979) *A Compendium of Alfalfa Diseases.* St Paul, Minnesota.

Humphreys, L.R. and Partridge, I.J. (1995) *A Guide to Better Pastures for the Tropics and Sub-tropics.* Tocal College, Patterson, NSW, Australia.

Stace, H.M. and Edye, L.A. (eds) (1984) *The Biology and Agronomy of* Stylosanthus. Academic Press, Sydney.

Wang, Y.R. and Hampton, J.G. (1991) Seed vigour and storage of 'Grasslands Pawera' red clover. *Plant Varieties and Seeds* 4, 61–66.

11

Leguminosae – Pulse Crops

The following species are dealt with in this chapter:

- *Pisum sativum* L. sensu lato – pea;
- *Vicia faba* L. – broad bean, field bean, faba bean, tick bean;
- *Phaseolus vulgaris* L. – French bean, haricot bean, kidney bean, snap bean;
- *Phaseolus lunatus* L. – Lima bean, sieva bean;
- *Vigna* spp.;
- *Lens culinaris* Medic. syn. *Lens esculenta* Moench. – lentil, split pea;
- *Cajanus cajan* (L.) Millsp. – pigeon pea, dahl, Congo pea, arhar;
- *Cicer arietinum* L. – chickpea, gram, Bengal gram, garbanzo bean;
- *Lupinus* spp. – the lupins; and
- *Lablab purpureus* (L) Sweet. syn. *L. niger* Medich., *L. vulgaris* Savi., *Dolichos lablab* L. – lablab bean, hyacinth bean, bovanist bean, seim bean, lubia bean, Indian bean, black-seeded kidney-bean, Egyptian kidney-bean.

***Pisum sativum* L. *sensu lato* (Pea)**

Origins and types

It is thought that the pea originates from Ethiopia, around the Mediterranean basin and central Asia with a secondary centre in the Near East, although there is no conclusive evidence of this (Davies, 1984). There is evidence that the crop has been cultivated for several thousand years in Europe. The modern cultivars have been developed from material originally introduced into Africa, China, Europe, India and North America from southwest Asia. This early distribution over a wide area is generally believed to be the reason for the present diversity of types. Peas are now widely cultivated in temperate regions and as a cool season crop in the tropics, especially at the higher altitudes, but mainly in northern Europe, parts of Russia and China and the northwest USA.

The pea crop has received increased attention since the mid-1950s; as a result of marketing the semi-mature seeds as a frozen product in Europe and North America, the product has become an important source of out of season 'garden peas'. The types developed for this outlet have the maximum number of pods developing for a single once-over harvest.

Some authorities have recognized two species, *Pisum arvense* L. (field peas) and *Pisum sativum* L. (garden pea). A further subdivision of *P. sativum* has also been used, i.e. *Pisum sativum* var. *macrocarpon* Ser. (edible pod peas) and *Pisum sativum* var. *humile* Poir (early dwarf peas in which the pods are lined with a characteristic membrane). However, all the types are now generally considered to be within the species *P. sativum*, which is divided into four main groups:

1. Vining peas – harvested while the seeds are still tender and used for immediate freezing, or canning and marketed as 'garden peas'.
2. Picking peas – grown as a horticultural crop for harvesting as a fresh vegetable, also harvested while the seeds are still tender and immature. Some cultivars in this group have tender pods lacking the pod's parchment layer; these types are consumed complete with immature seeds (frequently referred to as 'mange tout' or 'snap peas').
3. Combining peas – harvested for their dry seeds (in some areas or circumstances of small scale production they may be hand harvested). This group has two main uses for human consumption, either for the production of canned 'processed peas' or as dried peas. The other agricultural purpose is for animal fodder.
4. Forage peas – used for animal grazing, silage or haymaking.

The cultivated pea is an important species, especially in temperate regions. The crop is a very important source of protein in the human diet in many temperate or cooler areas of the world including higher altitude tropical areas where it is cultivated as a cool-season crop.

A detailed Test Guideline 007 is obtainable from UPOV (see Appendix).

The minimum standards for different cultivars' trueness to type are generally stricter for the seed crop categories being produced for food production compared with the final multiplication for the fodder types.

Site

The seed production site should not have been used for pea production in the previous 2 years. In areas where the important soil-borne pathogens and nematodes prevail a pea-free cropping period of up to 5 years is recommended.

The majority of weed seeds are easily removed from the pea seed lot during processing. However, the seeds of wild oat species (*Avena fatua*, *A. ludoviciana* and *A. sterilis*) can be difficult to remove from the seed lot during processing. If the crop has been damaged by pea moth larvae (*Cydia nigricana*), the larvae leave burrows in the seeds in which the wild oat seeds and some other weed grasses tend to lodge.

Peas are grown on soils with pH 5.5–6.5. The ratio of fertilizers applied in the base dressing during seedbed preparations depends on the nutrient status and local customs but specialist pea seed producers tend to use a N:P:K ratio of 3:1:2, which is a higher nitrogen level than that normally used for the market crop. Work by Browning and George (1981a) has indicated that increased seed yield can be obtained with relatively high levels of N and P. In addition, seeds produced from the higher N and P mother-plant regimes were found on analysis to contain higher levels of N and P. However, the seeds from the higher nitrogen regimes in the same experiments were found to be less vigorous when subjected to the conductivity test (PGRO, 1978). This indicated that nutrient regimes for seed yield differ from those required to produce high-quality seed.

Peas are very susceptible to manganese deficiency, especially on wet soils with relatively high organic levels. The symptoms of the deficiency are brownish hollow centres to the cotyledon seen when unripe seeds are split open. The condition is commonly called 'marsh spot'. Manganese sulfate is included in base dressings at a rate of 40–100 kg/ha where manganese deficiency is known to occur. Alternatively, manganese sulfate is applied as a foliar spray at a rate of 10 kg/ha in 200–1000 l of water, but this must be applied as soon as the symptoms are diagnosed otherwise it will be too late to improve seed quality and yield.

Another indication of low vigour in pea seeds is the incidence of 'hollow heart' and 'bleaching'. The former is characterized by a cavity of dead tissue on the adaxial surface of the cotyledons (Fig. 11.1) and the latter by a yellowing of the green seeds as a result of chlorophyll bleaching (Maguire *et al.*, 1973). It is generally accepted that both these conditions are caused by high temperatures during maturity or seed drying (Perry and Harrison, 1973). During the course of the investigations on mother-plant nutrition by Browning and George (1981b) it was shown that the low nitrogen and high

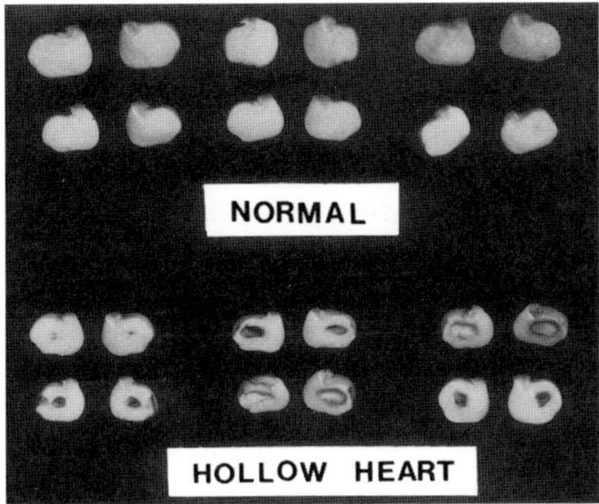

Fig. 11.1. Pea seeds with hollow heart compared with normal seeds.

phosphorus regime predisposed developing seeds to hollow heart whereas the high nitrogen regimes predisposed them to bleaching.

Pollination and isolation

Pea flowers are almost totally self-pollinated; fertilization occurs in the late bud stage before the flower is completely open.

Recommended isolation distances for peas are minimal in many countries and aim mainly to prevent admixture during harvesting. It is important that adjacent cultivars should be at least 20 m apart with the distance increased to at least 100 m for basic seed production.

Seed crop establishment

The winter-hardy cultivars are sown in the autumn, otherwise pea seeds are direct drilled into the field relatively early in the spring. The sowing rate per unit area is calculated to give a plant population of approximately 75/m^2; the sowing rate for the leafless types is 100/m^2.

The seed is generally sown in rows 12 cm apart. However, when either a high selection level is known to be required or the crop cannot be sprayed with specified pesticides, as with organic seed production, the inter-row distances can be increased up to 20 cm.

Some North American seed producers use a sowing rate of up to 250 kg/ha and a much shorter distance between rows for the final multiplication stage. The higher plant densities reduce the time span of pod maturity but increase the difficulty of roguing.

The pea crop has a moisture-sensitive stage from the start of anthesis until petal fall. This moisture-sensitive stage is very evident regardless of the amount of available water before or after flowering. Moisture deficits prior to anthesis only affect the weight of haulm produced, but shortage of water during pod growth can also affect yield.

The general practice in temperate regions is, therefore, to ensure that adequate irrigation is available during and after flowering, whereas in arid areas and dry seasons in the tropics sufficient water should also be given to ensure satisfactory plant growth before flowering commences.

Roguing

The seed crops are usually rogued at the following stages:

1. After emergence (when plants are approximately 15 cm high): remove the taller off-types. For basic seed production particular attention is given to checking that foliage, including stipules, is typical of the cultivar.
2. Flowering: remove early flowering plants from late-flowering cultivars, check flower colour and remove any plants with flowers that are not true to type. Check flower number per node.
3. Pods formed: check that pod shape, size and outer colour are typical of the cultivar; rogue out late-flowering plants, also non- or low-yielding plants.

Harvesting

The harvesting operation normally commences when the majority of pods have become parchment-like. By this stage the maturity of individual seeds is sufficiently advanced for them not to be adversely affected by subsequent drying. Biddle and King (1977) showed that the seed quality is reduced if seeds are harvested when their moisture content is above 30–36%. However, mature seed with a moisture content of 12% or less is subject to mechanical damage.

In areas where the seeds dry sufficiently on the plants the crop can be harvested by direct combining. A relatively low drum speed is used and care taken to avoid mechanical damage to the seeds.

Desiccants are used by some pea seed producers in areas where the pre-harvest drying-off of the plant material is relatively slow. They are applied when the crop starts to senesce and the lowest pods have turned pale brown and parchment-like. The usual rate of application, depending on product, is 3l in 100l of water sprayed on at the stage when random samples of seeds have a moisture content of 40%.

Crops that have been treated with a desiccant are either combined direct when the moisture content of seeds is approximately 28% or cut into windrows 4–5 days following application of the desiccant.

If combining is not possible due to unavailability of machines or the production scale is too small, an alternative is to cut the crop when the earliest pea pods have dried to the parchment stage and the foliage is starting to dry off; this is characterized by a reduction in the intensity of green in the leaves and haulm. Sample pods should contain fully developed seeds that are firm, taste starchy and are readily detachable from their pods.

Windrowed crops are usually turned, although each time the windrows are turned some seed is likely to be lost by shattering. An alternative to windrows is to cut and stack the crop on to tripods. This increases the rate of drying and reduces the time in which the seeds can deteriorate as a result of pathogen activity, but the method has a relatively high labour content. Material from windrows or tripods is threshed when the seed lot's moisture content is approximately 30%, using a thresher with rubber-covered tines.

Drying

Pea seed lots with their moisture contents above 14% should be further dried down to this level for storage. The overall drying rate should be relatively slow; those seed lots with a moisture content at 25% or above should have an initial air drying temperature of 5°C above the ambient temperature. For continuous-flow driers the air temperature should not exceed 43°C; this temperature should be reduced to below 38°C if the seeds' moisture content is 25% or higher.

Seed processing

Peas can be successfully processed on an air/screen cleaner. Colour sorters are used to remove discoloured seeds from the seed lot. Any seeds which have cracks or burrows from pea moth, weevils or mechanical damage can be largely removed by a needle drum separator or gravity tables. Large scale pea seed-processing plants should be designed to minimize the height that seeds have to fall during processing; augers should be avoided in pea processing plants.

Pests and pathogens

The main seed-borne pests and pathogens of *P. sativum* are listed in Table 11.1.

Vicia faba L. (Broad Bean, Field Bean, Faba Bean, Tick Bean)

Origins and types

Broad beans are believed to have originated and been initially domesticated in the Near East and gradually spread westwards along the Mediterranean coasts to Spain. Although China is a major producer of this crop the species arrived via the silk route comparatively recently.

The faba beans are cultivated for several reasons. There are several distinct types whose earlier classification has been abandoned following the fact that seed sizes of the modern cultivars no longer fit into the earlier defined groups. Although the current broad bean cultivars do not fit into discreet groups

Table 11.1. The main seed-borne pests and pathogens of *Pisum sativum*; these may also be transmitted to the crop by other vectors.

Pest or pathogen	Common names
Ascochyta spp., including *Ascochyta pisi*	Leaf and pod spot
Cladosporium cladosporioides f.sp. *pisicola*	White mould, leaf and stem spot
Colletotrichum pisi	Anthracnose
Erysiphe pisi	Powdery mildew
Fusarium oxysporum f.sp. *pisi*	American wilt (race 1), near wilt (race 2)
Fusarium solani	Root rot
Mycosphaerella pinodes	Foot rot, leaf spot, blight and black spot
Peronospora viciae	Downy mildew
Phoma medicaginis var. *pinodella*	Foot rot and collar rot
Pleospora herbarum	Foot rot
Rhizoctonia solani	Damping-off, stem rot
Sclerotinia sclerotiorum	Stem rot
Pseudomonas syringae pv. *phaseolicola*	Halo blight
Pseudomonas syringae pv. *pisi*	Bacterial blight
Xanthomonas rubefaciens	Purple spot
PEBV	Pea early browning virus
PEMV	Pea enation mosaic virus
PMiMV	Pea mild mosaic virus
PSbMV	Pea seed-borne mosaic virus, pea false leaf roll virus, fizzle top
BYMV	Bean yellow mosaic virus
Ditylenchus dipsaci	Pea eelworm

they can be grouped according to their main production purposes:

1. *Broad beans* harvested while the immature seeds are tender to be used as either a fresh vegetable or processed for freezing or canning. In some areas these are regarded as a pulse crop.
2. *Field beans* harvested when the seeds are mature; the dried seed is used as a stock feed. The broad beans produced for this purpose may be further subdivided as 'tic beans', which have small seeds, and 'horse beans', which have large seeds.
3. *Faba beans* are generally field beans in which the whole plant is harvested and used as silage or hay stock feed.

A detailed Test Guideline 206 is obtainable from UPOV (see Appendix).

Site

The majority of broad bean cultivars require long days for the acceleration of flower initiation although the earliest flowering cultivars do not. There is also evidence that vernalization at temperatures below 14°C will also accelerate flowering.

Soils for broad beans should have a pH of 6.5 with adequate available calcium otherwise a suitable liming material must be applied during soil preparations. The nutrient ratio applied during the final stages of seedbed preparations is an N:P:K ratio of 1:1:1. It is particularly important not to apply excessive nitrogen, especially for autumn-sown crops. This species does not tolerate waterlogged conditions.

Isolation

Broad beans are self-compatible but both self- and cross-pollination occur. Insect activity is responsible for up to 30% of cross-pollination. However, as the season advances some bees obtain nectar by making a hole at the base of the flowers, thus reducing the incidence of cross-pollination. High rainfall during anthesis can reduce or even prevent pollination by insects, which can result in a significant reduction of seed yield (Reisch, 1952).

Most authorities stipulate a minimum isolation distance of 300 m between broad bean crops. However, because of the relatively high incidence of cross-pollination, isolation distances should be at least 1000 m, especially for stock seed production.

Seed crop establishment

The long-pod group of cultivars which have some frost resistance are sown in the late autumn but in areas with severe winters sowings are made in early spring as with the other types of broad bean. Seeds are sown at the rate of 150 kg/ha in double rows 25 cm apart with 70 cm between the rows.

Broad beans are very responsive to irrigation applied during anthesis. Brouwer (1949, 1959) found that dry soil at this stage had the greatest adverse effect on yield while irrigation before the onset of anthesis had relatively little advantage even under dry conditions. However, Jones (1963) found that seed yield was to some extent dependent on the plants having a relatively high growth rate before flower development. It is therefore considered preferable to ensure that the plants receive sufficient water during early development and also from the start of anthesis.

Roguing stages

1. Before the start of anthesis, check plant habit vigour, height and tiller number, stipules, presence or absence of melanin spot, incidence of seed-borne pathogens, remove infected plants.
2. At start of anthesis, check general plant morphology, flower colour, including wing petals and standard, incidence of seed-borne pathogens, remove infected plants.
3. When pods set, check shape and relative length of pods, pose of pods.

Harvesting and processing

Signs indicating that a broad bean seed crop is ready for harvest include the pods becoming relatively dry with a loss of sponginess; this is preceded by a general blackening of the pods. The optimum seed moisture content at harvest is 16–20%. The crop is either cut by hand and tied in bundles or mown by machine. Bundling and stooking is necessary in northern Europe although not in the drier seed production areas such as the Middle East and North Africa. In areas where the seed is very ripe it is advisable to prevent or reduce shattering by combining early in the day while dew is still on the crop.

The harvested material is threshed when dry at a drum speed of 250 rpm. Care must be taken to adjust the concave setting according to the seed size of the cultivar to minimize mechanical damage to the seeds.

Drying

This is as described above for peas.

Pests and pathogens

The main seed-borne pests and pathogens of *Vicia faba* are listed in Table 11.2.

Phaseolus spp.

- Phaseolus vulgaris L. – French bean, snap bean, green bean, kidney bean, haricot bean, dwarf bean; and
- *Phaseolus lunatus* L. – Lima bean, sieva bean, butter bean.

Origins and types

Phaseolus vulgaris, which originated in North America, is cultivated throughout the temperate, tropical and subtropical areas of the world. It is largely cultivated in temperate areas for the immature green pods and is an

Table 11.2. The main seed-borne pests and pathogens of *Vicia faba*; these may also be transmitted to the crop by other vectors.

Pest or pathogen	Common names
Ascochyta fabae	Leaf and pod spot
Botrytis cinerea	Chocolate spot, grey mould
Botrytis fabae	Chocolate spot
Colletotrichum lindemuthianum	Anthracnose
Fusarium spp.	Foot rots, wilts
Pleospora herbarum	Net blotch
Uromyces viciae-fabae	Rust
Pseudomonas fabae	Stem blight
BYMV	Bean yellow mosaic virus
ClYMV	Broad bean mild mosaic virus (possibly caused by a strain of clover yellow mosaic virus)
BBSV	Broad bean stain virus
EAMV	Echtes ackerbohnenmosaik virus
Ditylenchus dipsaci	Stem eelworm

important crop for processing; the immature green pods are canned or frozen. In warmer regions the species is mainly cultivated for either dried beans or the dried beans are used in a range of ways including the preparation of canned navy or baked beans; there is also green bean production in the warmer regions during the cool seasons.

Specific cultivars have been developed for each of these processes and 'stringless' types have also been selected, which are suitable for the fresh crop and processing. The majority of cultivars are bush types but there are also climbing ('pole') types which are frequently referred to as 'climbing French bean'.

A detailed Test Guideline 012 is obtainable from UPOV (see Appendix).

Site

Phaseolus vulgaris tolerates slightly acid soil conditions; soils with a pH 5.4–6.5 can be used successfully. The general N:P:K ratio applied during seedbed preparation is 1:2:2. Gavras (1981) investigated the effect of mother plant mineral nutrition on seed yield and quality and found that different ratios of nitrogen and phosphorus were required for relatively high seed yield and for seed quality. Browning *et al.* (1982) reported the importance of a correct balance between nitrogen and phosphorus available to the plant for the production of high vigour seeds.

Pollination and isolation

Most cultivars of the bush and climbing types are day-neutral although there are some short-day cultivars in each group. The physiology of *Phaseolus* beans, including induction of flowering, pollination and pod growth, has been reviewed by Davis (1997).

The flowers are self-compatible and are predominantly self-pollinated although some cross-pollination occurs. The degree of cross-pollination probably increases in the tropics where insect activity, including thrips, is greater (Drijfhout, 1981).

Many seed authorities stipulate a minimum distance of 3m between a seed crop and any other crop of *Phaseolus vulgaris*. However, 50m for seed crops in the final multiplication stage and 150m for a crop intended for basic seed are preferable minimum isolation distances.

Seed crop establishment

Seed of the bush types is sown in the late spring when the possibility of frost has passed and soil temperatures have risen sufficiently for satisfactory germination and plant growth. In some specialized areas in North America, notably Michigan where length of season allows double cropping; seeds of early-maturing cultivars are drilled into the stubble of a previous barley crop or immediately after the stubble is ploughed in. When this technique is adopted seeds of early-maturing cultivars are drilled in early July.

The sowing rate depends on the relative seed size of the cultivar but is approximately 100kg/ha for the bush types and 50kg/ha for the climbing types. Bush cultivars are sown in rows 45–90cm apart, according to cultivar and stage of multiplication. The climbing cultivars are sown in rows 90–120 cm apart. In some tropical areas cane frames or supports are erected, although generally they are supported by strings.

Literature reviewed by Salter and Goode (1967) indicated that water shortages during anthesis and pod development seriously affected bean yield. Work in European temperate regions showed that irrigation applied before the start of anthesis only increased vegetative growth. However, as discussed by Davis (1997), early moisture stress, at the two trifoliate leaf stage, can have a detrimental effect on growth and affect flower initiation resulting in uneven crop maturity. There are many bean production areas where early and sustained irrigation is necessary to obtain a satisfactory crop.

Roguing

1. Before flowering: check plant habit, vigour and height according to type; check foliage, leaf shape and colour.
2. At start of flowering: check plant vigour and flower colour; remove plants showing symptoms of seed-borne pathogens.
3. Seed set and first pods formed: check pod characters, including shape and colour; remove late-flowering off-types and plants showing symptoms of seed-borne pathogens.

Harvesting

The dwarf or bush types are generally considered to be ready for a once-over harvest when the earliest pods are dry and parchment-like and the remainder of the pods have turned yellow. The seeds' moisture content at the time of harvesting should be between 20 and 25%; seed maturity is confirmed by opening sample pods, where the seeds should be fully developed with a mealy texture. Under good growing conditions the flowers tend to set until relatively late in the season. This leads to a loss from 'shattering' of the earliest mature seeds if harvesting is delayed. Smith (1955) examined the effects of stage at harvesting on bean seed yield and quality and found that there was a reduction equivalent to 358 kg/ha in harvested seed when the crop was cut before the earliest pods were mature.

Drying

The plants are cut and placed in windrows for further drying before either combining or threshing; alternatively they are combined direct from the standing crop. The entire operation is planned to ensure both the minimum loss from shattering and the least possible mechanical damage to the seeds, which are especially susceptible to cotyledon cracking.

The climbing cultivars mature over a longer period than the bush types so they are harvested by hand on three or more successive occasions as the older pods mature, otherwise the plants are pulled out and dried off in windrows. Smaller quantities of seed are usually spread thinly on a drying floor, but not directly exposed to the sun; if seed is being warm air dried the air temperature should not exceed 38°C.

Seed processing

Small quantities of seed are threshed by hand to avoid subsequent loss due to mechanical damage; this is especially important with small seed lots of basic or stock seed. Large seed lots are threshed with a drum speed of 250–350 rpm at a concave clearing from 12–20 cm. The seeds' moisture content should not be too low or excessive mechanical damage occurs during machine threshing.

Wilson and McDonald (1992) have evaluated six systems for the threshing of *Phaseolus vulgaris* and found that the highest quality seed was produced by open flail threshing or hand shelling. They concluded that, where feasible, manual threshing methods are superior to mechanical methods for small seed lots. Manual threshing would be expected to minimize cotyledon cracking.

Colour sorters can be used to remove off-type or discoloured seed from the final seed lot if necessary. Bean seeds are very prone to attack by the bean beetle (*Bruchus rupimanus*), which bores holes in the seed; weed seeds can lodge in these holes and may prove difficult to remove from the seed lot during seed processing. Figure 11.2 illustrates bean seeds with weevil damage.

Pathogens

The main seed-borne pathogens of *Phaseolus* species are listed in Table 11.3 (except for those of *Phaseolus lunatus*, which are listed

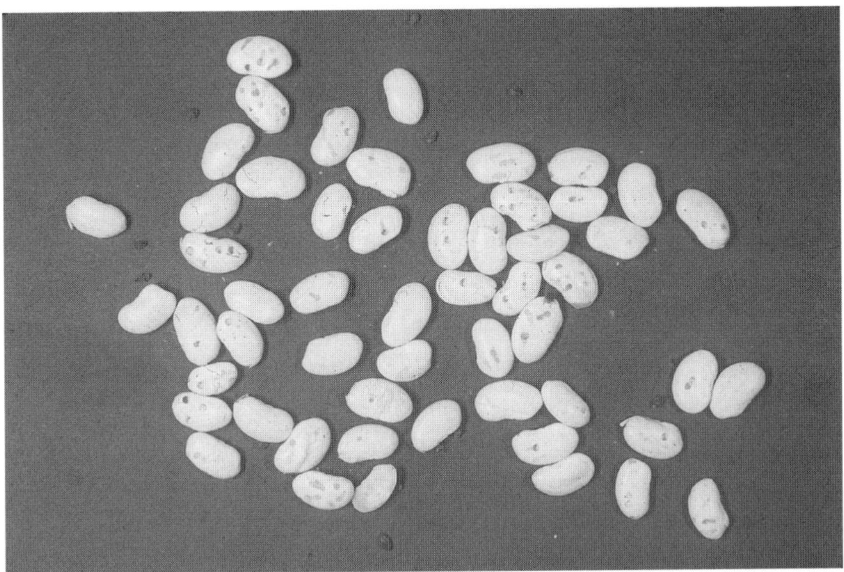

Fig. 11.2. Bean seeds with weevil burrows; the needle drum separator can remove seeds with this damage to upgrade the seed lot.

Table 11.3. The main seed-borne pathogens of *Phaseolus* spp. (except *P. lunatus*); these pathogens may also be transmitted to the crop by other vectors.

Pathogen	Common names
Alternaria alternata	Leaf spot
Ascochyta boltshauseri	Stem rot
Ascochyta phaseolorum and other *Ascochyta* species	Ascochyta leaf spots
Aspergillus sp.	'Baldheads' and 'snake-heads' of seedlings
Cercospora canescens	Leaf spot
Colletotrichum lindemuthianum	Anthracnose
Fusarium oxysporum f.sp. *phaseoli*	Yellows, wilt
Fusarium solani f.sp. *phaseoli*	Root rot
Phaeoisariopsis griseola	Angular leaf spot
Pleospora herbarum	Red nose, leaf spot
Rhizoctonia solani	Damping-off, stem canker
Sclerotinia sclerotiorum	Sclerotinia wilt, stem rot, watery soft rot and white mould
Pseudomonas syringae	Bacterial brown spot
Pseudomonas syringae pv. *phaseolicola*	Halo blight, grease spot
Xanthomonas campestris pv. *phaseoli*	Common bacterial blight, fuscous blight
BCMV	Bean common mosaic virus
CLRV	Cherry leaf roll virus
CMV	Cucumber mosaic virus

Table 11.4. The main seed-borne pathogens of *Phaseolus lunatus* (Lima bean); these pathogens may also be transmitted by other vectors.

Pathogen	Common name
Cladosporium herbarum	Cladosporum spot, stickiness
Colletotrichum truncatum	Stem anthracnose
Diaporthe phaseolorum and *Diaporthe phaseolorum* var. *sojae*	Pod and stem blight
Phoma exigua var. *diversispora*	
Phytophthora phaseoli	Downy mildew, collar rot
LGMV	Lima bean mosaic virus
BMV	Southern bean mosaic virus

in Table 11.4). These pathogens may also be transmitted to the crop by other vectors.

Phaseolus lunatus L. (Lima Bean, Sieva Bean, Butter Bean)

Origins and types

The Lima bean originates from South America with the main centres of origin in Mexico and Peru. It has been taken to parts of Asia and Africa where it is an important crop.

There are white seed types as well as black and speckled red, purple and brown. Some of the non-white types contain fairly toxic levels of hydrocyanic acid, which can normally be removed by changing the water at least once during the cooking time. The smaller seeded types are frequently referred to as sieva beans. There are both climbing and bush types of each of the different seed character selections. The bush types take 3 to 4 months from sowing to harvest while the climbing types take longer.

Site

The optimum soil pH range is 6.0–7.0. Although some types have reasonable drought resistance, a reasonably fertile and water-retentive soil is preferred.

Pollination and isolation

Although many growers regard Lima bean as self-pollinating, up to 18% cross-pollination has been reported (Purseglove, 1984). The isolation distance should therefore be 100 m.

Seed production

The general seed production requirements and techniques are as described for *Phaseolus vulgaris*.

Pests and pathogens

The main seed-borne pathogens of *Phaseolus lunatus* are listed in Table 11.4.

Vigna spp.

There are five main species of *Vigna*, which are tropical or subtropical pulse crops:

- *Vigna unguiculata* (L.)Walf. – cowpea, black-eyed pea, asparagus bean;
- *Vigna radiata* (L.) Wilczech var. *radiata*, syn. *Phaseolus aureus* – mung bean;
- *Vigna mungo* (L.) Hepper syn. *Phaseolus mungo* – black gram;
- *Vigna aconitifolia* (Jacq.) Marchal. syn. *Phaseolus aconitifolius* – mat bean, moth bean; and
- *Vigna angularis* (Willd.) Ohwi and Ohash – adzuki bean.

Vigna unguiculata (L.) Walf. (Cowpea)

Origins and types

Cowpeas are generally thought to have originated and become domesticated in tropical Africa where there are still several wild types. It is generally accepted that they were taken to Egypt from where the species was later taken to other parts of the Mediterranean and subsequently reached Asia. It is thought that the vegetable types were developed after its arrival in India. Material was taken to the West Indies and later to the USA where it has undergone considerable further development.

The main characters used to classify cowpea cultivars are:

1. The shape of the terminal leaf, plant hairiness.
2. Pod: angle of attachment, pigmentation, curvature, length.
3. Seed: shape, external texture, pigment pattern associated with the hilum.

Site

The minimum duration between cowpea seed crops is 1 year. However, some seed quality authorities may require a longer break, especially for the more elite seed classes. The break period should also take into account the presence of soil-borne pathogens including *Fusarim oxysporum* (wilt) and *Xanthomonas vignicola* (bacterial blight), which may require a significantly longer number of years' rest from cowpeas and related crops. Many seed producers programme this crop in rotations with other non-related species such as rice, maize or sorghum, depending on the local environment. The optimum soil pH is within the range of 5.6–6.5. The usual pre-sowing nutrient application is a base dressing of N:P:K in the ratio of 0:1.5:1. Nitrogen is withheld during site preparation, but when subsequent post-emergence growth is weak a light nitrogenous top dressing can be applied; however, excess available nitrogen will jeopardize flower and seed production.

Pollination and isolation

Although the cowpea is generally considered to be predominantly self-pollinated there can be some cross-pollination by a wide range of insects. The isolation distance is usually a minimum of 500 m. If local regulations allow closer distances between seed crops then a barrier crop of an unrelated species such as rice, sorghum or maize can minimize the contamination from foreign pollen.

Seed crop establishment

Furrow irrigation is preferable in the dry areas where seed production may be required, although care must be taken with the transfer of water to the furrows in order to avoid wetting the vines. Irrigation is beneficial during anthesis.

The stock seed is generally sown at the rate of 35 kg/ha although some adjustment for the sowing rate may depend on the cultivar's relative seed size. The seed rows of the dwarf cultivars are 30–60 cm apart with 15 cm between plants within the rows. The dwarf and intermediate cultivars are 'stopped' to produce bushier plants; this is done by pinching out the growing points when the plants reach a height of approximately 15 cm. The climbing cultivars are provided with 2 m-high supports and sown at 15 cm stations in rows 75 cm apart. Weed control is important for all morphological types in the early stages of plant establishment.

Roguing

There are three roguing stages:

1. Vegetative stage: routine checks to remove morphological off-types and plants with virus symptoms.
2. Seed pods nearing maturity: confirm external pod characters are according to the cultivar's description. Continue to remove plants displaying seed-borne virus symptoms.
3. Seeds in pods nearing maturity: confirm ripening seed characters are according to the cultivar's description.

Harvesting

Seed maturity of both the climbing and bush types is indicated by the external pods showing yellow-brown or brown coloration. The pods of the climbing cultivars are hand harvested when individual pods reach maturity. Generally the same concept can be applied to the shorter cultivars but those with even crop ripening can be cut and further dried in windrows. Hand harvesting is more productive for those cultivars with a wider maturity stage although these too can be cut, further ripened and dried in windrows, especially where there is a shortage of hand labour.

Threshing, drying, cleaning and storage

Those crops which have been left in windrows can be picked up by combine or otherwise carted to a stationary thresher. The drum speed should be adjusted so that mechanical damage to the seed is minimized, this is usually maintained at 800 rpm. The uniform feeding of threshers reduces the incidence of cotyledon cracking.

The maximum seed moisture content for storage is 13%. Further drying can be achieved on a drying floor with the air temperature not exceeding 35°C. The seed lot's moisture content should be further reduced to 11.5% for vapour-proof storage. The extracted seed is cleaned and upgraded as described for peas (*Pisum sativum*).

Pests and pathogens

The important seed-borne pathogens of *Vigna* spp. are listed in Table 11.5.

Vigna radiata (L.) Wilczech var. *radiata*, syn. *Phaseolus aureus* (Mung Bean, Green Gram)

Origins and types

The mung bean is of high economic and culinary importance in Asia where it is used for the production of 'bean shoots' or 'bean sprouts' and also as a pulse. In some areas it is also grown as a cover crop for livestock fodder; the immature pods are used as a vegetable. It is cultivated in the tropics and grows well in a temperature range of 30–35°C. The species is drought tolerant and is less successful in high rainfall areas.

Cultivar classification

- Seed testa colour of ripe seed: yellow, green/yellow, light green, dark green, brown, multicoloured.
- Growth type: indeterminate, determinate.

There are no climbing types, although *V. radiata* is more erect than other *Vigna* species.

Site

There should be a minimum of a 1 year gap between related crops; some seed authorities may require a longer break, especially for the higher seed classes. Any previous history of soil-borne pathogens should also be taken into account when selecting the site.

The crop tolerates a soil pH of 5.5–6.5. Depending on the soil nutrient status a general base fertilizer with an N:P:K ratio of 1:8:5 is incorporated during seedbed preparation. A nitrogenous top dressing can be applied if the crop is weak during the establishment stage, but no nitrogen should be applied after the start of anthesis.

Pollination and isolation

Vigna radiata is generally regarded as self-pollinating. Normally there is only a need for a gap of 3 m between crops to avoid admixture at sowing and harvesting.

Seed crop establishment

The seed sowing rate is 10 kg/ha in rows 20 cm apart with the aim to establish plants at 8 cm stations within each row. Particular attention should be given to the control of the pathogen *Macrophomina phaseolina* (ashy stem blight) during seed production, especially when the seed crop is to be used for the production of bean shoots.

Pests and pathogens

The main seed-borne pathogens of the *Vigna* spp. are listed in Table 11.5.

Vigna mungo (L.) Hepper. syn. *Phaseolus mungo* (Black Gram, Urd)

Origins and types

This *Vigna* species is an economically and nutritionally important crop on the Indian subcontinent, where it is mainly cultivated as a fodder crop.

The cultivars are classified according to their seed size and testa characteristics as either larger with black pigmentation, or smaller with olive-green pigmentation (Purseglove, 1985). The mature pods are black.

Seed crop establishment and seed production

The production and subsequent operations are as described above for mung bean.

Vigna angularis (Willd.) Ohwi and Ohash syn. *Phaseolus angularis* (Willd.) Wight (Adzuki Bean)

Origins and types

The adzuki bean is a cultigen not found in the wild form. It originates in Asia and is generally believed to have been cultivated in China from where it was taken to Nepal and Japan. It is an important food crop in Japan. The species has been introduced as a commercial crop to the southern USA, South America and some areas of Africa, in particular the Congo. This species is cultivated as a grain legume, vegetable and as a fodder crop.

There are dwarf determinate and dwarf semi-determinate types; there are also indeterminate climbing types. Mature seed characters are included in the classification described and used at the International Institute of Tropical Agriculture (IITA), Ibadan, Nigeria. The characters referred to are the seeds' eye patterns, including shape of the pigment pattern around the hilum and eye pigmentation.

Site

The adzuki bean is a relatively slow germinator and does not compete very successfully with weed species in its early stages. It is therefore important that there is an effective weed control programme in advance of the seed crop and during its early stages of establishment.

The crop requires a soil with a pH of approximately 6.0. Soils with a lower pH should be dressed with a liming material one crop in advance of the adzuki seed crop.

Pollination and isolation

This species is predominantly cross-fertilized; isolation distances should be at least 50 m.

Seed crop establishment and seed production

Seed production of adzuki bean is as described above for green gram (mung bean).

Harvesting

The crop is usually harvestable approximately 17 weeks from sowing. However, the seed pods shatter very readily when nearing maturity and care should be taken with harvest timings and procedures.

Table 11.5. The main seed-borne pathogens of *Vigna* spp.; these pathogens may also be transmitted by other vectors.

Pathogen	Common names
Colletotrichum spp.	Anthracnose
Macrophomina phaseolina	Ashy stem blight
Xanthomonas campestris pv. *vignicola*	Bacterial spot, blight, canker
BCMV	Bean common mosaic virus
CpABMV	Cowpea aphid borne mosaic virus
CpGBVB	Cowpea green vein banding mosaic virus
CPMoV	Cowpea mosaic virus
SBMV	Southern bean mosaic virus (cowpea strain)
TMV	Tobacco mosaic virus
ULCV	Urdbean leaf crinkle virus

Drying and seed processing

These are as described for green gram.

Pests and pathogens

The main seed-borne pathogens of *Vigna* spp. are listed in Table 11.5.

Other Pulses

Lens culinaris Medic. syn. *Lens esculenta* Moench. (Lentil, Split Pea)

Origins and types

According to Ladizinsky (1999), the genetic stock from which lentil has been derived can be found in three lines collected in Turkey, northern Syria and southern Syria. Lentil cultivation was originally closely associated with wheat and barley. Langer and Hill (1991) describe two seed types: (i) the large-seeded cultivars with large flat pods bearing flattened seeds 6–9 mm in diameter with yellow to orange cotyledons and large white or occasionally blue flowers; and (ii) the small-seeded cultivars having seeds 3–6 mm in diameter with small convex pods, convex cotyledons and small violet-blue or pink flowers.

The crop is a cool season crop often cultivated as a winter crop in temperate regions and is widely cultivated in the Mediterranean region, Near East, India and France. Although not a tropical crop it is also cultivated at higher altitudes in the tropics during the dry season.

A detailed Test Guideline 210 is obtainable from UPOV (see Appendix).

Site

Lentil will succeed on a wide range of soil types but will not succeed in wet conditions. Sites known to have *Orobanche* spp. present should be avoided. Lentil should not immediately follow *Lathyrus* or *Vicia* species as it is difficult to differentiate between the young plants of each of the three species during early roguing. Lentils respond well to phosphate fertilizers applied during site preparation.

Pollination and isolation

The lentil is self-fertilized although some cross-pollination is believed to occur. The minimum isolation distance is 10 m.

Seed crop establishment

Seed crops of lentils are easier to manage when grown as a row crop. The seed is sown at the rate of 8–10 kg/ha in rows 50 cm apart with seed at approximately 20 cm apart within the rows for the early cultivars. The later types are sown at the rate of 12–15 kg/ha in rows 120 cm apart with the seeds approximately 60 cm apart within the rows.

Roguing

Off-types and rogues should be removed during cultural operations, otherwise the main roguing is done at the start of anthesis, when flower colour can be checked.

Harvesting, drying and seed processing

The seed crop is usually hand-pulled at early seed ripening and further dried prior to threshing.

The drying and seed crop processing of lentils are as described for cowpeas.

Seed-borne pests and pathogens

The main seed-borne pathogens of lentil are listed in Table 11.6.

Table 11.6. The main seed-borne pathogens of lentil; these pathogens may also be transmitted by other vectors.

Pathogen	Common names
Ascochyta lentis	Ascochyta blight
Botrytis cinerea	Botrytis, grey mould
Fusarium oxysporum	Fusarium wilt
PsbMV	Pea seed-borne mosaic virus

Cajanus cajan (L.) Millsp. (Pigeon Pea, Dahl, Congo Pea, Arhar)

This species is a very important crop for subsistence farmers especially in India, southern and eastern Africa, Central America and the Caribbean. The main agronomic value of pigeon pea is in its tolerance of poor soils. Pigeon peas crop well in harsh conditions although they do not tolerate waterlogging or significant shade.

Origin and types

There has been much discussion during the 20th century as to the origin of pigeon pea; however, it is now generally accepted that it originated in India, was taken subsequently to Malaysia and from there to East Africa and later to the Americas. The species is not very successful in the humid tropics.

There are two groups of cultivars recognized in India:

1. The Tur group, *Cajanus cajan* var. *flavus*: the cultivars in this group mature early, are short stemmed with yellow flowers, glabrous pods, usually light green with three seeds.
2. The Arhar group, *Cajanus cajan* var. *bicolor*: the cultivars in this group mature later, are bushy perennials, the inflorescence standards are either striped or blotched with red or purple pigmentation; the pods are pubescent with red or purple splotches or stripes, containing four or five seeds.

Recent plant breeding activities by the International Crops Research Institute for the Semi-Arid Tropics (ICRISAT) have improved the crop's productivity. These include the production of short-duration cultivars which mature in 100–150 days; the development and introduction of the shorter morphological types has increased the yield per unit area. ICRISAT has provided material for Indian private- and public-sector breeders for the production of hybrid cultivars.

Site

The site should not have produced a leguminous crop in the previous 2 years. Care should be taken to ensure that volunteer legume species are controlled during the previous 2 year's cropping and the final plot preparations. The optimum soil pH is 6.5; if the crop is to be grown on soils tending to be acid then a base dressing of a liming material should be incorporated during soil preparations. Pigeon pea crops respond well to a single base dressing of phosphate fertilizers, usually applied at the rate of 40–60 kg/ha. This species will tolerate drought.

Pollination and isolation

Pigeon pea is cross-pollinated; the recommended minimum isolation distance is 400 m for the higher seed classes.

Seed crop establishment

Although some pigeon pea cultivars can produce an economic crop when maintained for more than 1 year the seed crop is normally produced as an annual. The pigeon pea seedling emerges better from a rough seedbed rather than one prepared with a fine tilth which would tend to form a soil cap. The sowing rate is 50–70 kg/ha for the small-seeded cultivars and 75–120 kg/ha for larger seeded cultivars. The seed is sown in rows 30–120 cm apart, the distance depending on the cultivar's size at maturity and plant morphology.

Roguing

The most preferable stage for roguing is when the crop has started to flower; however, for certified or higher seed classes the following sequence of roguing stages is recommended:

1. Late vegetative: plant morphology according to cultivar.
2. Start of flowering: relative time from emergence to start of anthesis; flower colour, ivory, yellow, orange, red or purple.
3. Pod colour: green, purple, multicoloured or striped.

Harvesting

Large areas of the seed crop are normally cut and left in windrows until dry and then either picked up with a combine or taken to a stationary thresher. Small plots can be hand pulled and further dried in the field prior to being gathered for threshing.

Drying

Seed moisture content should be further reduced when it is above 14% moisture content. This can be achieved on a ventilated floor or by a continuous-flow or batch dryer; the forced air temperature should not exceed 35° C.

Seed processing

As described for cowpeas.

Seed-borne pathogens

The main seed-borne pathogens of pigeon pea are listed in Table 11.7.

Cicer arietinum L. (Chickpea, Gram, Bengal Gram, Garbanzo Bean)

Origin and types

The chickpea originates from central and western Asia (van der Maesen, 1987). It is an important cool-season crop in India and is widely cultivated in the Middle East and Iran. It is often grown as a winter crop close to the Mediterranean. Many of the wild types are sub-shrubs but the cultivated cultivars are annuals. The crop tolerates drought but does not do well in the humid tropics. The chickpea has several culinary uses including as a pulse when split, whole in dahls or chickpea flour.

Two types are recognized in the Indian subcontinent:

1. 'Desi', which are the small coloured-seeded cultivars and landraces.
2. 'Kabuli' has larger and white seeds.

Table 11.7. The main seed-borne pathogens of pigeon pea; these pathogens may also be transmitted by other vectors.

Pathogen	Common name
Alternaria alternata, *A. tenuissima*	Alternaria blight
Aspergillus flavus	A storage disease, also a producer of aflatoxin
Cochliobolus lunatus	Cochliobolus leaf spot
Colletotrichum cajani	Anthracnose
Fusarium oxysporum	Fusarium wilt
Fusarium moniliforme	Fusarium wilt
Phoma sp.	Canker
PPSMV	Pigeon pea sterility mosaic virus
TBRV	Tomato black ring virus

A detailed Test Guideline 143 is obtainable from UPOV (see Appendix).

Site

There should be a break of at least 2 years between any previous leguminous crop and chickpea. The crop is suitable for heavy soils provided the drainage is satisfactory. The emerging seedlings can tolerate a coarse seedbed. The crop responds well to both phosphorus at the rate of 50–70 kg/ha and potassium at the rate of 45 kg/ha incorporated as a base dressing; where the soil's nitrogen status is low the crop will respond to a light dressing of nitrogen soon after seedling emergence, at a rate not exceeding 15–20 kg/ha.

Pollination and isolation

The chickpea is considered to be completely self-pollinating, although some authorities may require isolation for the higher seed classes. Otherwise only a short distance of 10–20 m is sufficient to avoid admixture at sowing or harvesting.

Seed crop establishment

The smaller seeded types are sown at the rate of 50–70 kg/ha, larger seeded types are sown at the rate of 75–120 kg/ha. The seeds are sown in rows 30–120 cm apart depending on the vigour and morphology of the cultivar.

Roguing

Roguing stages:

1. Vegetative stage, before the start of anthesis: stem and leaf pigmentation, absent or present; number of leaflets per leaf, few (3–9), many (more than 13).
2. As soon as anthesis can be observed: plant height: width and stem length; time of flowering: early, mid or late season; inflorescence: number of flowers per peduncle; flower colour: blue, pink or white; relative number of secondary shoots: few or numerous.

Other characteristics can be observed from the mature pods and seed; although these are useful for cultivar descriptions, they cannot normally be used for roguing as the seedcoat character is maternal in origin and also it is too late to rogue out significant off-types as their mother plants have already contributed to the gene pool of the seed lot to be harvested.

Harvesting, drying and seed processing

Small areas of seed crops can be cut or pulled by hand. Following drying the material can be hand threshed. Larger areas can either be combined or cut and left to dry in windrows before picking up. If the seed is further upgraded by a forced-air drier the air temperature should not exceed 35°C. Other postharvest operations are as described for pigeon peas or where referred to cowpeas.

Pests and pathogens

The main seed-borne pathogens of chickpea are listed in Table 11.8.

Table 11.8. The main seed-borne pathogens of chickpea; these pathogens may also be transmitted by other vectors.

Pathogen	Common name
Alternaria brassicicola	Alternaria blight
Ascochyta rabiei	Anthracnose, ascochytosis
Aspergillus flavus	Collar rot
Botrytis cinerea	Grey mould
Fusarium oxysporum f.sp. *ciceris*	Fusarium wilt
Fusarium solani	Fusarium foot rot
Fusarium spp.	Root rot and wilt complex
Pleospora herbarum	Pleospora leaf spot
Stemphylium sarciniforme	Leaf spot

Lupinus spp. (The Lupins)

Origins and types

The lupins are natives of North and South America and parts of the Mediterranean basin. The New World centres of diversity are central Mexico and California. The genus is cultivated for grain, forage, soil amelioration and also for its amenity value. It is thought that their decline as domestic crops came about partly as a result of domestication of other leguminous species and also because the seeds had a relatively high alkaloid level. Plant breeders have been developing lines that have lower alkaloid levels to the extent that there are both 'bitter' and 'sweet' lines in each of the species listed below.

The main cultivated *Lupinus* species are:

- *Lupinus albus* L. – white lupin believed to have been developed from indigenous species in the Balkans;
- *Lupinus angustifolius* L. – blue lupin, narrow leaf lupin;
- *Lupinus cosentinii* Guss. – Western Australian common blue or sandplain lupin. This species is cultivated as a fodder crop in Western Australia;
- *Lupinus luteus* L. – yellow lupin;
- *Lupinus mutabilis* Sweet. – pearl lupin, tarwi. This species originated in the Andes, Peru and Colombia, and has the highest oil level of all the domesticated *Lupinus* spp. Mainly cultivated in South America, and an important parent in breeding programmes.

A detailed Test Guideline 066 is obtainable from UPOV (see Appendix).

Site

The genus is not frost hardy, although some cultivars require vernalization from temperatures above freezing. Except for *L. albus* and *L. angustifolius*, the other species do not tolerate alkaline conditions. Lupin seeds have a long dormancy period and are very susceptible to soil-borne pathogens including *Fusarium* spp. and *Sclerotinia* spp., therefore there should be a break of 4 years between lupin crops and any other leguminous species.

Pollination and isolation

Cross-pollination can be relatively high within the species. In addition to the usual problems of maintaining cultivar purity there is also the question of crossing or contamination between the bitter and sweet cultivars. The sweet cultivars generally have 0.05% bitter principle. Some cultivars have marker genes that are very beneficial when roguing, selecting or inspecting seed crops in the field. The placing of hive bee colonies in or on the edge of a crop will improve seed set and also help to maintain the required pollination level within the plot.

Recommended isolation distances for fields exceeding 2 ha for seed crops intended for further multiplication is 100 m and 50 m for grain crops. The distances for fields of less than 2 ha are 200 m and 100 m, respectively.

Seed crop establishment

The rate of germination can be slow in low soil temperatures and also in wet soils. The optimum plant density is 50–60 plants/m^2; these high plant densities encourage the production of a single dominant inflorescence per plant, which assists seed crop uniformity at harvest. The high plant populations also assist weed suppression within the lupin crop. Lupin stock seed is sown in rows 10–18 cm apart.

Roguing stages

1. Start of anthesis: time of flowering, main flower colour, stem colour; pattern of inflorescence: determinate, semi-determinate, intermediate.
2. Seed setting: pod colour; seed characters. The following are useful for cultivar descriptions and classification: colour pattern, main colour, eye colour (around hilum), eye size, seed shape.

Harvesting, drying and processing

The crop is usually cut and left in windrows for further drying in the field; it can then either be picked up with a combine or carted to a stationary thresher. Some large-scale producers use a normal header-type harvester. Shattering can be reduced in arid conditions by harvesting early in the day and while the dew is still on the crop. Small areas are hand pulled, field dried and then threshed.

Pests and pathogens

The main seed-borne pathogens of lupins are listed in Table 11.9.

Lablab purpureus (L) Sweet. syn. *L. niger* Medich., *L. vulgaris* Savi., *Dolichos lablab* L. (Lablab Bean, Hyacinth Bean, Bovanist Bean, Seim Bean, Lubia Bean, Indian Bean, Black-seeded Kidney-bean, Egyptian Kidney-bean)

Origins and types

This species is believed to have originated in India and possibly some adjacent areas in Asia. It succeeds in arid climates and is an important crop of subsistence farmers, especially in southern India and some other tropical areas including the Sudan where it has become established as an important break crop in the Gezira cotton areas, traditionally following cotton and sorghum, prior to 1 year's fallow.

The lablab bean is a weak perennial, mainly cultivated as a dried pulse crop and to a small extent for the young immature green pods, which are used as a vegetable. The haulm, either green or dried, is used as a livestock feed or forage. This species is also used for green manure, silage and as a cover crop. According to Purseglove (1984) there are two botanical types: *Dolichos lablab* var. *lablab* and *D. lablab* var. *lignosus*, which has pods that are more truncated than those of *D. l.* var. *lablab*; however, Smartt (1976) suggests that these two types probably represent extremes of a more or less continuously variable population. There are named cultivars, but these are mainly based on seed colour differences between different lines.

Site

A well-drained site with relatively low available nitrogen is preferable for the seed

Table 11.9. The main seed-borne pathogens of lupin species; these pathogens may also be transmitted by other vectors.

Pathogen	Common names
Botrytis cinerea	Grey mould, seedling rot, branch canker
Colletotrichum spp.	Anthracnose
Diaporthe woodii	Phomopsis stem blight
Fusarium oxysporum	Wilt
Gibberella avenacea	Damping off
Glomerella cingulata	Anthracnose
Pleiochaeta setose	Brown spot
Sclerotinia sclerotiorum	Sclerotinia rot
Stemphylium spp.	Stemphylium leaf spot complex
Verticillium albo-atrum	Wilt
Erwinia sp.	Black stem
Pseudomonas spp.	Stem rot
BYMV	Bean yellow mosaic virus
BYMV Narrow leaf strain	Bean yellow mosaic virus, narrow leaf strain
CMV	Cucumber mosaic virus, browning
PSV	Peanut stunt virus

crop otherwise the haulm and leaf growth is excessive. The optimum soil pH range is 5.5–6.0. Lablab requires a weed-free seedbed as it does not compete well during early crop establishment, although once established it competes very well with the weed population.

Pollination and isolation

Lablab bean is both self- and cross-pollinating, but for seed production purposes it should be regarded as cross-pollinated and with 200–400 m isolation. Separate seed stocks should be isolated by a distance of 500 m.

Seed crop establishment

Some selections of lablab are weak perennials; although the seed crop is frequently grown as an annual some farmers keep the plants to produce a crop in the second year. Although this is satisfactory for the ware crops it is not good policy for seed production. The stock seed is sown at a rate of 30–60 kg/ha in rows 1 m apart. In small-scale seed production it may be produced between wide rows of maize or sorghum. The seeds should be treated with a cowpea strain (CB 756) of *Rhizobium* if being produced on a new site.

Roguing

The seed crop is rogued at the start of anthesis to remove diseased plants, especially those with symptoms of seed-borne diseases or displaying off-type morphological characters, including flower colour, according to the seed stock's description.

Harvesting

The plants are hand pulled and field dried in windrows prior to seed extraction.

Drying

The moisture content should be reduced to 13% for short-term storage or down to 11% for periods of up to a year and stored in vapour-proof containers.

Seed processing

The seed is usually hand threshed, which produces a cleaner seed lot. Otherwise small threshers can be used.

Pests and pathogens

The main seed-borne pathogens of lablab bean are listed in Table 11.10.

Table 11.10. The main seed-borne pathogens of lablab bean; these pathogens may also be transmitted by other vectors.

Pathogen	Common name
Colletotrichum lindemuthianum	Anthracnose
Trichothecium roseum	Anthracnose
Xanthomonas campestris pv. *phaseoli*	Bacterial blight

Further Reading

Biddle, A.J. and Catlin, N.D. (2007) *Pests, Diseases and Disorders of Peas and Beans: a Colour Handbook.* Academic Press, London and New York.

Brunt, A.A. (1996) *Viruses of Plants.* CAB International, Wallingford, UK.

Cockbain, A.J. (1983) Virus and virus-like diseases of *Vicia faba* L. In: Hebblethwaite, P.D. (ed.) *The Faba Bean* (Vicia faba *L.*). Butterworths, London, pp.421–462.

Hebblethwaite, P.D. (1983) *The Faba Bean (Vicia faba L.*). Butterworths, London.

Pilbeam, C. and Hebblethwaite, P. (1994) The faba bean. *The Biologist* 41(4), 169–172.

Smartt, J. (1976) *Tropical Pulses.* Longman, London.

van der Maesen, L.H.G. (1987) Origin, history and taxonomy of chickpea. In: Saxena, M.C. and Singh, K.B. (eds) *The Chickpea.* CAB International, Wallingford, UK.

12

Edible Oilseed Crops

This chapter contains those important agricultural crops cultivated for the production of the main edible vegetable oils and includes oilseed crop species of the following families:

- *Cruciferae*;
- *Asteraceae* (formerly *Compositae*);
- *Leguminosae*;
- *Pedaliaceae*.

Oilseed Crop Species in *Cruciferae*

- *Brassica napus* L. var. *oleifera* – Swede rape, hungry gap kale;
- *Brassica rapa* L. (formerly *B. campestris*) – turnip rape, Polish rape;
- *Brassica juncea* (L.) Czern. and Coss. – brown mustard (brown-seeded cultivars), yellow or Oriental mustard (yellow-seeded cultivars);
- *Sinapis alba* L. (syn. *B. hirta* Moench.) – white mustard.

Origins and types

The oilseed-producing brassicas are important crops in the temperate areas of Europe and North America and are also cultivated in Asia. *Brassica napus* was an ancient crop in the Mediterranean basin. *Brassica juncea* is also cultivated as a pot herb and leafy vegetable, especially in the Indian subcontinent. The work by Brown *et al.* (1997) has indicated that *Sinapis alba*, an oilseed crop in some areas, especially parts of the Indian subcontinent but a common weed in other geographical areas, has some important and potentially useful traits; these include tolerance to some insect pests of *Brassica* spp. and also good drought- and high temperature tolerance; these characters could therefore be of importance in future breeding and cultivar development of *Brassica* cultivars. There is also interest in some of these species for the production of biofuels.

Detailed Test Guideline 036 for rape seed and Test Guideline 185 for turnip rape are obtainable from UPOV (see Appendix).

Site

The oilseed-producing brassicas dehisce very readily and their seeds are capable of a long viability when left in the soil. There should therefore be a period of at least 3 years between related oilseed crops being grown on the site; this should be extended to 5 years for the production of basic or earlier seed generations. The overall exclusion of previous cropping within 5 years with any *Brassica* species is preferable, with the exclusion including vegetable and forage brassicas. This practice will reduce the incidence of seed admixture and soil-borne pests and pathogens.

The most important weed species whose seeds are difficult and time consuming to eliminate from the harvested seed crop are *Sinapis arvensis* (charlot), *Raphanus raphanistrum* (wild radish), *Brassica nigra* (black mustard) and *Galium aparine* (cleavers). The problems and control measures of weeds in oilseed crops have been discussed by Lutman (1991).

Pollination and isolation

The oilseed rapes and mustard are cross-pollinated and should have a minimum isolation distance of 500 m when the crop is to be used for further seed production, otherwise 200 m is usually considered sufficient.

Crop establishment

The winter cultivars are sown in time to establish the crop prior to the start of winter. The spring cultivars respond well to early sowing unless they are grown in areas where climatic conditions are likely to deter early crop establishment. For winter rape, NIAB (1996) suggests sowing 120 seeds/m^2 to provide a stand of 80 plants/m^2 by the end of winter. The spring-sown cultivars have smaller seeds than the winter types; NIAB suggests a sowing rate of 150 seeds/m^2.

When producing hybrid cultivars that have been developed using cytoplasmic male sterility the ratio of female:male parents from 3:1 to 7:1 is generally adopted. The hybrid seed crop is also surrounded by four or five rows of the male parent. The male parent plants should be removed prior to harvesting the hybrid seed crop from the remaining female parents. The harvested seed will also be male sterile and should be mixed with a pollinator seed stock in the ratio of 4:1.

Roguing

The roguing operations for this crop group are extremely important because the cultivar purity must be maintained to very high standards; this is for two reasons:

1. The crops are primarily cultivated for the oil extracted from the seed and the low level of erucic acid is essential.
2. The residue after oil extraction is an important source of protein for stockfeed and some other purposes; contamination with other cultivars, or cross-compatible species, can result in toxic residues in the meal.

There are very specific analytical requirements for each of these products, which require high cultivar purity. Plant breeders have produced cultivars with safe and reduced levels of erucic acid, as well as cultivars with reduced levels of glucosinolates in the meal, which would otherwise be harmful to livestock. The condiment manufacturers require mustards with a specific level of pungency in their products.

The initial roguing should be made while the crop is still in the vegetative stage, well before the start of anthesis. Satisfactory isolation should be confirmed at this time; confirm that isolation is according to specified requirements of the seed certification agency. A second roguing should be at the start of anthesis, when the flower colour should be checked. At this time flower colour should correspond to the species and appropriate cultivar description.

Harvesting

The optimum stage for harvesting and securing the best potential seed yield is usually within a period of 6 days, otherwise seed is lost through shattering. *Brassica* seeds are shed from their fruiting bodies very readily when ripe and the timing of harvesting operations is critical in order to secure the maximum yield.

There are three possible systems for a satisfactory seed harvest:

1. Cut and leave in windrows; when sufficiently dry, pick up with a combine and thresh when the seed moisture content has reduced down to between 35% and 20%. The plants are starting to turn yellow at this stage and the seed is becoming a darker colour and can be marked with a thumb nail. This method is also suitable when cutting by hand in which case the cut crop is tied into bundles to dry in the field, or with small volumes of material it may be dried in an airy structure during inclement weather. Figure 12.1 illustrates a cut *Brassica* crop drying in windrows.

2. The standing crop is desiccated at a slightly later stage than cutting into windrows and subsequently directly combined when the seed moisture content is down to 14%. This can be a very useful method with a very weedy crop although there is a danger that desiccation can damage the crop seed and reduce its viability.

3. Direct combining when the crop is at a stage estimated to give the maximum seed yield. The optimum plant stage for this is when the plants have changed colour, the seed is dark in colour, the seed moisture content at this stage should be 14% or less and the seed hard. A pick-up attachment should be used with combines if combining directly from the windrows. It is advisable to seal combines to reduce the risk of seed loss.

Drying

The high oil content in the seeds of these crops leads to rapid seed deterioration if the moisture content has not been reduced to 9% or less prior to up to 6 months' storage. The moisture content should be reduced to less than 7% at the start of longer storage periods. All samples intended for longer term storage should be less than 5% moisture content.

The relatively small seeds of these crops pack together more tightly than larger seeded species. Therefore it is important when on-floor drying that the depth of seed should not exceed 1m; however, less than 1 m can cause pockets of wet seed to form between the air inlets (Weiss, 1983). In areas where field drying is risky as a result of inclement autumn weather the plants can be pulled and dried in an airy structure, as shown in Fig. 12.2.

Ward *et al.* (1985) recommend a maximum air temperature applied for drying floors of 5°C above ambient; the drying temperature

Fig. 12.1. A cut brassica seed crop drying in windrows.

Fig. 12.2. A brassica seed crop drying off in an unheated glasshouse in England.

for ventilated bins is recommended as 7°C above ambient. Cereal continuous-flow driers may be used but care should be taken to ensure that seed is not lost through gaps. The seed depth on flat-bed driers should be adjusted so that neither restriction of airflow or 'bubbling' can occur.

Seed processing

This crop group can normally be processed satisfactorily with an air/screen cleaner.

Pathogens

The main seed-borne pathogens of *Brassica* spp. are listed in Table 12.1.

The main seed-borne pathogen of *Sinapis alba* is *Alternaria brassicae*.

Table 12.1. The main seed-borne pathogens of *Brassica* spp.; these pathogens may also be transmitted to the crop by other vectors.

Pathogen	Common names
Albugo candida	White blister
Alternaria brassicae	Grey leaf spot, leaf spot
Alternaria brassicicola	Black spot, wirestem
Erysiphe sp.	Powdery mildew
Leptosphaeria maculans	Dry rot, blackleg, black rot, phoma
Mycosphaerella brassicicola	Black ring spot
Plasmodiophora brassicae	Club root
Rhizoctonia solani	Root rot
Sclerotinia sclerotiorum	Watery soft rot, drop, white blight
Xanthomonas campestris pv. *campestris*	Black rot
TuYMV	Turnip yellow mosaic virus

Oilseed Crop Species in *Asteraceae* (formerly *Compositae*)

- *Helianthus annuus* L. – sunflower;
- *Carthamus tinctorius* L. – safflower.

Helianthus annuus L. (Sunflower)

Origins and types

The sunflower originates from the western USA and is thought to have subsequently spread to Central America as a camp-following species; it was subsequently introduced to South America and later taken to Europe. It has become established as an important oilseed crop in Russia in addition to other areas of Europe.

The wild *H. annuus* is multi-headed and it is believed that the single-headed (monocephalic) cultivars of modern production were originally selected from it. Figure 12.3 illustrates the wild multi-headed type, which may be present as off-types or rogues in the seed stock of a monocephalic cultivar. There has also been the development of hybrids as well as cultivars of specific heights from very short to very tall. The shorter types have been developed for combine harvesting while the medium height cultivars provide easier harvesting by hand.

A detailed Test Guideline 081 for sunflower is obtainable from UPOV (see Appendix).

Site

Fields known to be infected with *Orobanche* (broomrape) should not be used for this crop for a minimum of 4 years. *Cuscuta* (dodder) can be a threat to the crop although there are

Fig. 12.3. An illustration of the wild *Helianthus annuus*, on the left, compared to a monocephalic cultivar. Figure reproduced by kind permission of J.F. Hancock, from *Plant Evolution*. Copyright CABI 2004.

efficient herbicides for its control in sunflower. The most difficult weeds are the indigenous sunflower species in North America, which cross-pollinate with the higher oil-yielding cultivars. A break of 1 year is generally considered sufficient for the elimination of volunteer sunflowers produced from fallen seed.

Pollination and isolation

The inflorescences of this crop are normally cross-pollinated by a wide range of insects including bees; wind pollination is also a strong possibility.

The minimum isolation distance of a commercial seed crop from any other sunflower crop is 400 m, although this distance should be increased when the crop is grown for a higher seed class; some authorities require isolation distances to be 800–1500 m according to the seed class. The greater distances should be adopted when producing inbred parental material or hybrids.

Seed crop establishment

The open-pollinated seed crop is cultivated in the same way as for a commercial oil crop. Plants are sown in rows 55–100 cm apart, with 15–30 cm between plants within the rows. The spacing depends on the vigour and height of the cultivar. The seed is sown at a rate of 4–10 kg/ha. Graded seed is recommended when precision drilling.

The sowing pattern for the production of hybrid seed is usually specified by the breeder or supplier of the parental lines. Vannozzi (1987) recommends that the proportions of female cytoplasmic male sterile (CMS) to male pollinator line should be in a ratio of 2:1 to 7:1 (female to male). He proposes plant populations from 45,000 to 67,000/ha. Langer and Hill (1991) suggest a density of 30,000/ha in dry soil regimes and 60,000/ha when irrigation is available.

Before embarking on hybrid seed production it is essential to confirm that the two proposed parents will flower at the same time; seed producers use the term 'nicking' for the simultaneous flowering of the parent lines. In some cases the sowing times of each have to be modified in order to achieve a satisfactory 'nick'.

Roguing

The inflorescences turn to the east following the development of the ray florets; therefore it is advisable to have the plant rows oriented north–south so that the flower heads can be inspected by looking westwards.

Harvesting

Seed-head maturity is indicated when the underside of an inflorescence has turned from yellow to brown; the seeds will have become hard and difficult to mark with a thumb nail by this stage. The cut heads' moisture content should be approximately 12% when it is planned to combine or thresh the seed heads. In some arid areas there is a risk that seed shedding will occur when the heads become too dry. Desiccants can be applied to the standing crop with care so that there is no adverse effect on germination potential.

The whole seed head is harvested. A selective harvest can be made when the seed heads are cut off by hand from smaller areas or plots; the seed heads are cut with the minimum length of stalk. The seed heads can also be hand-cut and dried in the field unless rains are likely.

Drying

All seed harvested at or above 12% moisture content should be reduced to 10% or lower. The drying process for seed harvested at 20% and above must commence immediately otherwise potential germination will be significantly reduced. The harvested seed is very likely to contain large quantities of other seed head and plant material, usually pieces of flower head and stalk; these may have a very high moisture content and it is recommended that they are removed by screen and aspiration on a pre-cleaner as soon as possible so as to reduce the overall drying requirement. In some cases these materials and/or off-colour seeds can be hand sorted where there is plentiful hand labour available as seen in Fig. 12.4.

The seed can be floor or ventilated bin dried with ambient air heating. If heated air is used its temperature should not exceed 50°C.

Seed processing

This can normally be done efficiently with an air/screen cleaner. Following this the seed can be upgraded by either an indent cylinder or a grader. The seed lot can be size graded if required.

The main seed-borne pathogens

The main seed-borne pathogens of sunflower are listed in Table 12.2.

Carthamus tinctorius L. (Safflower)

Origins and types

This species is thought to have originated in the Middle and Near East. Its florets were sewn onto papyrus or cloth wrapped around mummies; more recently it has been used as a substitute for saffron and the production of an orange vegetable dye (safflower carmine); the dried inflorescences can still be purchased in the Iranian and Egyptian bazaars for use as dyes and food colourants. It is an important oilseed crop in several geographical areas

Fig. 12.4. Hand sorting sunflower seeds for discoloured seeds. By kind permission of The Real IPM Company (K) Ltd (www.realipm.com), the copyright holders.

Table 12.2. The main seed-borne pathogens of sunflower; these pathogens may also be transmitted by other vectors.

Pathogen	Common name
Alternaria alternata	Leaf spot (responsible for aflatoxins on seed)
Alternaria helianthi	Leaf blight, leaf spot
Alternaria sesami	Alternaria leaf spot
Botrytis cinerea	Grey mould
Leptosphaeria linquistii	
Macrophomina phaseolina	Charcoal rot
Plasmopara halstedii	Downy mildew
Sclerotinia sclerotiorum	Wilt, white stem rot, stem rot
SuMV	Sunflower rugose mosaic virus

including the western USA, Australia, Iran and India. The expressed meal or cake is a valuable livestock feed.

Cultivars are basically classified according to their 'spininess' according to spines on the outer involucral bracts, leaf morphology, oil and dye content. There are cultivars with known levels of oleic and linoleic fatty acid contents.

A detailed Test Guideline 134 is obtainable from UPOV (see Appendix).

Site

The roots of this species can penetrate its substrate down to approximately 4 m, thereby reaching nutrients and water not available to many other crops. Safflower is generally produced in rainfed areas and tolerates drought and dry winds. Volunteer plants are not normally a problem as the seed is usually expected to have a short dormancy and volunteers can be controlled during site preparation. Therefore only a 1-year break is normally required between seed crops and a previous safflower crop, although some national regulations may require a longer period. However, if grown in rotation with wheat or barley the volunteers from these cereals can be a problem. The previous herbicide history of the proposed site should be checked as safflower is intolerant of their residues. Previous crop infections of *Sclerotinia* mould should be carefully recorded as safflower is very susceptible to this pathogen.

Seed crop establishment

The seed crop is managed in the same way as for the commercial oil crop. However, because of the species' proliferation of spines, walkways should be left when sowing to allow for roguing and certification inspections.

The young plants form an early rosette following seedling establishment but weed control is important in the early stages as the young crop does not compete well with local weed species before forming a rosette. Safflower responds well to the fertilizers and manures with a high nitrogen content.

Seed is sown in rows at a seed rate of between 15 and 40 kg/ha. The higher plant densities result in a higher proportion of primary stems per unit area; the primary stems produce the best quality seed.

Roguing

The main stage for roguing is as soon as the flower colour can be confirmed as according to the cultivar.

Harvesting

Safflower generally requires 4–5 months from sowing to seed harvest depending on cultivar and location. The seed is hard when ripe, and can be easily squeezed out of the seed heads; the whole plant is brittle by this stage and the bracts will have become a darker colour. Despite the crop's spines and the discomforts of working with it, the crop does not lodge or have a reputation for seed shedding in the field. Therefore in most production areas or seasons it is advantageous to wait for the seed moisture content to reduce in the field to approximately 8%. When this is not possible the crop is combined at 10–12% moisture content and immediately further dried down to 5–8%. Small areas can be hand cut and dried in windrows prior to threshing. Exceptionally weedy crops are best cut and windrowed when seed moisture content is 20–25%. In this case picking up from the windrows should wait until the seed moisture content has reduced to less than 8%. Wet weather can stimulate germination in the seed head.

Safflower seed is weak hulled, therefore care should be taken during its harvest; threshing with low cylinder and drum speeds should reduce the risk of seed damage.

Drying

It is generally emphasized that seed is best dried on the plant while still in the field. When harvested at 8% moisture content or higher immediate further drying is essential. Bin driers are the better method for this crop otherwise continuous-flow driers with the air temperature controlled at 40°C, or lower, are suitable.

Seed for the following season should be stored with its moisture content at 8% or lower; although smaller seed samples are best maintained in refrigerated stores.

Seed processing

The seed can be processed and upgraded by an air/screen cleaner.

Pests and pathogens

The main seed-borne pathogens of safflower are listed in Table 12.3.

Table 12.3. The main seed-borne pathogens of safflower; these pathogens may also be transmitted by other vectors.

Pathogen	Common name
Alternaria carthami	Leaf spot, crown necrosis
Cercospora carthami	Leaf spot
Fusarium oxysporum f.sp. *carthami*	Fusarium wilt
Puccinia carthami	Safflower rust
Verticillium albo-atrum	Verticillium wilt
Verticillium dahliae	Verticillium wilt
A strain of lettuce mosaic virus, LMV	Safflower mosaic virus

Oilseed crop species in *Leguminosae*

Arachis hypogea L. (Groundnut, Peanut)

Origins and types

This crop species makes a very significant contribution to human nutrition. The seeds contain 43–55% oil and 25–28% protein. The crop is widely produced in the tropics and tropical America. The species originates from Brazil. It has become an important food crop in Asia and tropical Africa. The meal remaining after oil extraction is used as a very high protein livestock feed.

According to Tindall (1983) there are two types of *A. hypogea*:

1. Virginia type, which is branching; some forms have true prostrate runners or spreading bunches; flowering only occurs on lateral stems, alternating with vegetative shoots.
2. Spanish-Valencia type; this type has upright branches, flowering on upper nodes, stems angular in early growth stages, older stems hollow and cylindrical.

A detailed Test Guideline 093 is obtainable from UPOV (see Appendix).

Site

Groundnuts are very susceptible to weed competition immediately following seedling emergence, a weed-free seedbed is therefore recommended. *Amaranthus palmeri* (Palmer amaranth) is a particular problem in parts of the USA. Seeds remaining in the soil from previous crops soon germinate and can therefore be controlled during site preparations. Therefore a break of 1 year is sufficient between a previous groundnut crop and a seed crop, although some authorities may stipulate a longer period, especially for the higher seed classes. Proposed sowing sites should be as free as possible of the remains of previous crops in order to minimize stem rot.

The optimum soil pH is 6.0–6.5. Soils known to be low in available calcium should receive a dressing of lime or other calcareous material as calcium deficiency can lead to 'blind' pods resulting in low yields. The seeds are developed below ground, therefore the ideal soil is friable and deep. Seed producers usually produce the seed crop on a site with residual soil nutrients from a previous crop because groundnuts do not respond significantly to direct fertilizer applications; the seed is sensitive to fertilizers that are incorporated in the sowing substrate. The crop is not successful on heavy soils.

Pollination and isolation

Arachis hypogea is mainly self-fertilized, although some cross-pollination by bees occurs. Generally, only a 3 m wide strip is considered sufficient isolation to prevent admixture during sowing and harvesting.

Seed crop establishment

The seed crop is produced in the same way as for a commercial ware crop. The seed is sown in closely spaced rows; the close spacing assists the development of compact plants, which are easier to harvest.

The mother seed can be sown with precision drills; there are specific peanut plates available for plate drills. Seed can be sown in ridges, especially when irrigation methods dictate this; however, this can reduce the number of 'pegs' which the plant can bury. It is easier to harvest from ridges than from flat soil surfaces.

Roguing

The ideal stage for roguing or seed crop inspection is at the start of anthesis; this provides the best opportunity for the identification and roguing out of off-types.

Harvesting

The number of days from sowing to harvest will depend on the cultivar and climatic conditions; the general rule of thumb is:

- approximately 120–140 days from sowing for the Virginia types;
- approximately 90–100 days for the Spanish-Valencia types.

However, those cultivars that produce three or four seeds per pod require up to 180 days to maturity.

The foliage remains green in many cultivars at seed maturity; it is therefore necessary to sample a series of subterranean pods prior to harvesting so as to determine the timing of the overall harvest (Fig. 12.5). The pods become dark on the inside at maturity and the seeds become their final colour according to the cultivar. At this stage the seeds have 30–40% moisture content and the crop can be lifted.

Care must be taken in crop handling to avoid damaging the pods by splitting or bruising. The crop can be lifted either by hand forking or mechanically. Any mechanical damage can lead to subsequent splitting, which can result in reduced germination potential or serious fungal activity lethal to the seed.

The plants are left in the field to dry after they have been lifted. Ideally the crop is placed on poles or tripods, but this requires hand labour. Otherwise the harvested crop is left in windrows but the pods should not remain resting in the soil. Stack drying results in better quality seeds because each seed receives a similar amount of exposure to the drying elements compared with windrows where some are under the drying material while the top ones may be overexposed. The pods are removed from the haulm when the seed moisture content is at or just above 20%.

Drying

It is generally recommended that crops grown for seed production are de-hulled

Fig. 12.5. Checking maturity stage of a Nepalese farmer's groundnut crop.

as near to the time of sowing as possible. Therefore the pods are firstly dried prior to their storage. The seeds arriving at the store can be expected to have a moisture content of approximately 20%. Prior to storage they should be dried down to just below 10% to ensure that the seeds maintain the best possible condition and remain free of seed-borne pathogens and pests. The drying down should be carefully monitored – if dried too fast the shells become brittle; the required seed moisture content should not be exceeded. The air temperature during drying should be approximately 35°C and not exceed 38°C at any stage. The hessian sacks of pods can be left to dry outside in many of the warmer and drier areas where the seed crop is produced.

Seed processing

Shelling should be done carefully so as to minimize seed damage. The shelled seed should be further cleaned to remove any broken pods and debris to upgrade the final seed lot.

Pests and pathogens

The main seed-borne pathogens of groundnut are listed in Table 12.4.

Glycine max (L.) Merrill. (Soybean)

Origins and types

The soybean is believed to have evolved as a domesticated species in the eastern part of

Table 12.4. The main seed-borne pathogens of groundnut; these pathogens may also be transmitted by other vectors.

Pathogen	Common name
Aspergillus niger	Crown rot, collar rot
Macrophomina phaseolina	Root rot, stem rot
Mycosphaerella arachidis and *M. berkeleyi*	Tikka disease, black spot, early leaf spot, blotch, late leaf spot
Puccinia arachidis	Rust
Rhizoctonia solani	Damping off
PeMoV	Peanut mottle virus

north China; the cultivated types were subsequently taken to Korea, southern China, Japan and southern Asia.

According to production and export statistics soybean is the most important grain legume crop in *Leguminosae*. It is an important crop in the USA, Brazil and China and several other South American and Asian countries.

The crop and its manufactured products for human consumption are utilized in several ways including milk types, powders and curds. The mature seeds contain 20–23% oil and 39–45% meal depending on the cultivar. The meal provides a high protein stock feed, especially in the production of pork, poultry and their products. Some cultivars have been developed for use as a fresh green vegetable; the adoption and use of the vegetable soybean is still considered to be undeveloped, further increasing the importance of this cultivated species which may also be cultivated as a pulse crop.

The North American and Canadian classification places the cultivars in a sequence of 00, 0 and I to X; those in group 00 and 0 are adapted to the areas with the longest days. However, the North American groupings do not necessarily apply in other parts of the world because soybeans are also sensitive to temperature and moisture availability during anthesis. IBPGR (1984) gives five groups, which correspond to the 12 North American groups as listed in Table 12.5.

Site

Soils with a pH of 5.7–6.2 are suitable; those with a lower pH should receive a lime dressing during site preparation. The crop is not suitable for waterlogged soils.

Table 12.5. Soybean cultivar groupings according to the IBPGR and the North American schemes.

IBPGR Grouping	USA/Canadian Grouping
1	00, 0
2	I, II
3	III, IV
4	V, VI, VII
5	VIII, IX, X

The soybean crop is very susceptible to weed competition; the indigenous or wild grasses can be especially competitive especially in the seed crop's early stages. Subsequent problems during seed processing can also occur when there are leguminous volunteers of similar seed size in the crop. Therefore special attention should be paid to the site's previous cropping during the site selection. The seeds of previous soybean crops are short lived in the soil and do not present a problem as volunteers in a following crop, although national seed regulations may stipulate a longer gap between crops according to the seed class being produced.

Pollination and isolation

The flowers are self-pollinated; cross-pollination is regarded as less than 1%. Generally a minimum gap of 3 m is considered sufficient to avoid admixture during sowing and harvesting although national seed regulations may stipulate a greater distance for some seed classes.

Seed crop establishment

Seed should be inoculated with *Rhizobium japonicum* if soybean has not been grown on the site previously. This microorganism is specific for soybean and its presence is necessary for satisfactory levels of nitrogen-fixing bacteria, which are required for optimum growth. The seed is sown in rows 50–100 cm apart at a sowing rate to give a stand of one plant for each 2.5 cm of row; it is believed that the seed crop ripens more evenly at higher plant populations per unit area. The closer spacing encourages more compact plants, which can be helpful when harvesting. Seed can be sown with a precision drill or otherwise there are plate planters that can be fitted with a peanut plate. Some growers prefer to sow in ridges; in some cases this may be dictated by the irrigation system that may be required.

Weeds should be controlled, especially during crop establishment, because young soybean plants are weak competitors.

Roguing

The most important morphological characters can be observed during anthesis and again at crop maturity; therefore seed authorities may require two inspections for certified and higher class seed crops. The main roguing stage is at the start of anthesis.

Harvesting

The physiological maturity of the seed crop is indicated by yellowing of the seed; in many cultivars this occurs prior to any significant indication of leaf senescence. Pod shattering and seed shedding will occur from this stage onwards. An application of a fungicidal spray can be a useful protection for the maturing seeds at this stage when production is in the humid tropics.

Although small areas can be hand harvested, threshing should be done as soon as possible so as to remove the seed to satisfactory storage conditions. The crop is very suitable for direct combine harvesting; the optimum seed moisture content for this is 14%, although careful adjustment must be made when the seeds' moisture content is either greater or lower than this. The seeds are soft when above 14%, if at lower levels they become brittle and prone to cracking. The cylinder speed should be reduced to lessen the possibility of mechanical damage at the lower moisture levels.

Drying

Initially a seed moisture content of 14% or less is essential prior to storage; ideally 12% is better for safe storage. Soybean seed does not store well and even good quality seed will start to deteriorate after approximately 6 months, especially in open storage where neither temperature or humidity are controlled.

The drying air temperature should remain at 40°C or a little below. The drying rate should be gradual when the seed lot's initial moisture content is above 20%; the air temperature should also be reduced.

Seed processing

Soybean seed can normally be successfully cleaned on an air/screen cleaner; passing the seed lot through a gravity separator can be an additional upgrading.

Seed storage

For storage of up to 6 months the seed moisture content should not exceed 12%, and ideally be maintained at 10%. The seed lot's moisture content should be checked at intervals, and should be re-dried when moisture levels are at, or above the optimum level. The seed should be stored at a moisture content not exceeding 10% when kept in an air-conditioned store maintained at 60% humidity and 20°C.

Small quantities of high value seed lots can be stored in vapour-proof containers at 5°C.

Pests and pathogens

The main seed-borne pathogens and pests of soybean are listed in Table 12.6.

Oilseed Crop Species in *Pedaliaceae*

Sesamum indicum L. syn. *S. orientale* L. (Sesame, Simsim, Beniseed, Gingelly, Til)

Origins and types

Sesamum indicum has been recognized as a source of edible oil and seeds for thousands of years and has long been widely cultivated in parts of Africa, Asia and South America. It has become a commercial crop in North America; several countries continue to investigate its potential as a commercial crop, including India, China and Myanmar (formerly Burma). The pressed

Table 12.6. The main seed-borne pathogens and pests of soybean; these pathogens may also be transmitted by other vectors.

Pest or pathogen	Common name
Cercospora kikuchii	Purple blotch, purple speck, purple seed stain
Colletotrichum truncatum	Seedling blight, anthracnose
Diaporthe phaseolorum var. *sojae* and *Phomopsis* anamorphs	Stem canker, pod and stem blight
Fusarium spp.	Fusarium root rot, fusarium wilt, collar rot
Nematospora coryli	Yeast spots
Peronospora manshurica	Downy mildew
Pseudomonas syringae pv. *glyinea*	Bacterial blight
Pseudomonas syringae pv. *tabaci*	Wildfire
Xanthomonas campestris pv. *glycines*	Bacterial pustule
Heterodera glycines	Soybean cyst nematode
SMV	Soybean mosaic virus
TRSV	Tobacco ringspot virus, bud blight

remains from oil extraction are used as a stock feed. The seeds are also used for the preparation of halvah, which is a popular confection in the Middle East. The green leaves are used as a pot herb and for the preparation of soups in parts of Africa. There are some indigenous *Sesamum* species in Africa, including *S. alatum* and *S. radiatum*, which are collected by some households although they are not of commercial importance.

Cultivar description:

- long day or short day cultivar;
- season of sowing and season of seed production, branching habit;
- flower number in leaf axils; loculi per seed capsule: four, six or eight;
- external corolla pigmentation: white, pink, purple;
- seedcoat colour: ranging from white to black; generally the black-seeded types are thought to be undesirable, while the white seed types are considered to produce the best quality oil;
- rate of seed capsule ripening from the base of plant upwards;
- rate of capsule dehiscence.

Plant breeders have developed cultivars that are suitable for mechanized harvesting; this has been achieved by eliminating the dehiscent character. Other breeding improvements have included increases in seed yield and disease resistance. Because the flowers are predominantly self-pollinated local landraces can be improved especially for the stabilization of flower and seedcoat colour characters.

Site

The seed crop should not follow on from a previous sesame crop unless it was the same cultivar of a higher seed class. Neither should it follow any species with a similar-sized seed, otherwise there will be difficulty in separation of the crop seed from any admixture that has occurred from other crop volunteers, indigenous and wild sesame species. Some weed or volunteer species, such as *Sorghum* spp. and *Digitaria* spp. with similar sized seeds to sesame, should be controlled prior to or during seedbed preparation. Johnson grass should also be controlled because its seeds are extremely difficult to separate from those of sesame.

Site

The optimum soil pH is approximately 6.5. The crop thrives on relatively dry soils; however, as germination is sometimes slow irrigation may be required after sowing and also during the early stages of anthesis; the crop cannot tolerate waterlogging. This species does not succeed in the humid tropics.

Seed crop management

The developed sesame plant has a relatively long taproot and the plant can thrive fairly well under dry conditions. Generally seed is sown towards the end of the rainy season in parts of Asia and at the start of the rains in Africa.

Seed is usually sown in drills from 40 to 80 cm apart; the distance depends on available soil water for satisfactory germination. The normal sowing rate is 4 kg/ha. Commercial crops are often interplanted with other crops of local importance, but this is not normally advisable when producing a seed crop.

Pollination and isolation

Sesame is largely self-fertilized, although cross-pollination of up to 10% by insects should be anticipated. Recommended isolation distances vary from one country to another; 400 m is recommended between different cultivars, while the certification standard in some countries is 100 m for Foundation Seed and 50 m for Certified Seed.

Roguing

Roguing is first done at the start of anthesis with a second roguing when the seed capsules form. These stages are also suitable for seed-crop inspections, although the number of inspections may be increased by the authorities in some countries to include a pre-flowering stage.

Harvesting

It should be stressed that the seeds are delicate and easily cracked or otherwise damaged; therefore the seed crop should be handled with care to maintain potential storage life and germination. Unless non-shattering cultivars have been grown care must be taken to choose the best timing for the seed-crop harvest. Many growers with smaller farms cut and stook the crop while the seed capsules are still green but when the foliage starts to turn yellow. The seeds are then extracted by flailing on a clean smooth surface, plastic sheet or tarpaulin. Non-shattering cultivars can be cut when mature and left in windrows; however, where the crop has been produced in fields with a ridge and furrow system care must be taken to ensure that the cut material is left across the furrows and not in the furrows. The material can be directly combined from the windrows. Large-scale producers invariably grow non-shattering cultivars and are therefore able to mechanically harvest direct from the mature crop.

Seed processing

Threshers should be run at a relatively low speed with the concaves fully open. All seed-cleaning equipment should be sealed to reduce losses resulting from the seed's free-flowing character. Sesame seed is normally cleaned on an air/screen cleaner.

Drying

Cleaned seed is reduced to 8% moisture content at ambient RH prior to storage, or to 5% for immediate storage in vapour-proof containers.

Pests and pathogens

There are two important seed-borne pathogens of sesame, which may also be distributed by other vectors. The pathogens are: *Cercospora sesame* Zimm. (leaf spot) and *Alternaria sesami* (Kawamura) Mohanty and Behera (Alternaria leaf spot).

Further Reading

Oilseed crops in *Cruciferae*

Frauen, M. (1987) Technological and economic aspects of seed production of hybrid varieties of rape. In: Feistritzer, W.P. and Kelly, A.F. (eds) *Hybrid Production of Selected Cereal Oil and Vegetable Crops*. FAO, Rome, pp. 281–300.

Kimber, D. and McGregor, D.I. (eds) (1995) *Brassica Oilseeds Production and Utilization*. CAB International, Wallingford, UK.

Robbelon, G. (1987) Hybrid varieties of rape. In: Feistritzer, W.P. and Kelly, A.F. (eds) *Hybrid Production of Selected Cereal Oil and Vegetable Crops*. FAO, Rome, pp.133–148.

Sunflower

Nayar, N.M. (1976) Sesame. In: Simmonds, N.W. (ed.) *Evolution of Crop Plants*. Longman, London and New York, pp.231–233.

Purseglove, J.W. (1974) *Tropical Crops, Dicotyledons*. Longman, London, pp. 430–443.

Vannozzi, G.P. (1987) Technological and economic aspects of seed production of hybrid varieties of Sunflower. In: Feistritzer, W.P. and Kelly, A.F. (eds.) *Hybrid Production of Selected Cereal Oil and Vegetable Crops*. FAO, Rome, pp.253–279.

13

Forage Crops of *Cruciferae* and *Chenopodiaceae*

Cruciferae – Forage Crucifers

- *Brassica oleracea* L. convar. *acephala* – fodder kale;
- *Brassica oleracea* var. *capitata* L. – cabbage;
- *Brassica oleracea* var. *gongylodes* L. – kohlrabi;
- *Brassica napus* L. var. *napobrassica* (L.) Rchb. – swede, rutabaga;
- *Brassica rapa* L. (including *B. campestris* L.) – turnip; and
- *Raphanus sativus* L. var. *oleiferus* – fodder radish.

Flower initiation in most of the cole crops is dependent on a low-temperature stimulus. This vernalization requirement varies between the different types and between cultivars within the individual types. The physiology of the main vegetable cole crops with flower initiation have been reviewed by Wien and Wurr (1997).

Brassica oleracea (Fodder Kale, Cabbage, Kohlrabi)

Origin and types

In addition to the *Brassica* spp. listed above there are several important temperate vegetables that are subspecies of *B. oleracea*, including: Brussels sprout, the vegetable cabbages and savoys, cauliflowers (winter and summer types) and broccoli (sprouting and calabrese). All of these *B. oleracea* subspecies, including *B. oleracea*, which is the wild species from which they have all been selected and further developed by growers and seedsmen over the centuries, are cross-compatible. Seed production of the main vegetable crucifers has been described by George (2009). A description of the early history and domestication of *B. oleracea* and the development of the different *B. oleracea* types has been discussed by Thompson (1984). The cultivated *B. oleracea* varieties are often referred to collectively as cole crops.

It is thought that the Greeks were cultivating kales by 600 BC (Thompson, 1984). Early domestication would have included selecting parental plants with lower levels of glucosinolates in order to reduce bitterness in the crushed foliage.

There are two basic types of fodder kale in the traditional classification, 'Marrow stem' and 'Thousand head', although plant breeders have produced cultivars with improved desirability which do not clearly correspond to either of these types.

1. Marrow stem kale: this type has a long, thin and fleshy stem which livestock can eat; the plants have relatively long internodes; when subsequent bolting starts axillary

shoots emerge from the extension shoots. This type is comparatively tall.
2. Thousand head kale: the stems of this type are shorter and thinner and are both tough and less palatable to livestock. This type has short internodes and is much leafier as a result of axillary shoots originating early from the early plant growth stage. The two types are considered together, including the hybrid and triple-cross types.

The history and development of the main *cabbage* types has been discussed by Thompson (1984). Although widely grown in temperate regions there are some cultivars that have been developed for tropical and subtropical conditions. There are cultivars produced specifically for stockfeed but some of the culinary cultivars are also used.

A detailed Test Guideline 048 is obtainable from UPOV (see Appendix).

There are open-pollinated and F1 hybrid cultivars of *kohlrabi*, identified by:

- season and uses, suitable as stockfeed, resistance to early bolting;
- 'bulb' characters: colour; red, white, purple, extent of pigmentation if speckled;
- shape: globe or flat;
- internal quality: degree of fibre (if tendency to be fibrous); and
- foliage: short or tall, pose of leaves, relative length of petioles.

A detailed Test Guideline 065 is obtainable from UPOV (see Appendix). The UPOV description refers to the marketable part of the plant as the 'kohlrabi' while many seedsmen and growers refer to it as the 'bulb'; however, botanically it is a swollen stem.

Choice of site for *Brassica* seed crops

Crucifer seed has a long soil survival rate. Therefore there should be at least 5 years between cruciferous seed crops; no cruciferous crops should be grown on the proposed site for the 2 years prior to sowing the designated seed crop. All cruciferous weeds and volunteer crop species should be eliminated before they produce seed during this final 2-year preparation time.

The important weeds that should be excluded in the preparation are *Sinapis arvensis* (charlock), *Rumex* spp. (docks), *Chenopodium album* (fat hen), *Galium aparine* (cleavers), *Geranium dissectum* (cranesbill) and *Melandrium album* (white campion). This list of undesirable weed species includes those whose seeds present separation problems during brassica seed processing.

Fields with a history of the pathogen *Plasmodiophora brassicae* ('club root', also commonly known as 'finger and toe') should not be used, especially as this pathogen can be seed-borne in addition to being soil-borne. Other important seed-borne pathogens include *Alternaria brassicicola* (dark leaf spot) and *Phoma lingam* (canker).

The optimum pH for these three crops is 6.0–6.5 and any required liming material should be applied during soil preparation. Acid soil conditions not only affect the availability of micronutrients such as molybdenum, but also increase the incidence of club root (*Plasmodiophora brassicae* Woron.).

The N:P:K ratio of nutrients applied during preparation varies widely between different seed production areas, but the general recommendation is 1:2:2. High nitrogen levels result in 'soft' plants less able to withstand frost, and for this reason top dressings of nitrogen are only applied in the spring. The spring application also replaces nitrogen leached during the winter. Where deficiencies of available manganese, boron or molybdenum exist, suitable additions of the appropriate materials should be applied. A solution of sodium molybdate is applied to the young plants prior to planting out in areas where this is likely to occur. Boron deficiency can be responsible for 'hollow-stem' or corky patches on the stems of cole crops.

Most references in the literature relating to the irrigation of cole crops refer to their vegetative stage. Increasing the available water when the plants are in the vegetative stage will increase the amount of foliage produced but no work has reported on the effects of water availability on seed yield. Water stress in the cole crops increases the thickness of the waxy cuticle and produces plants with a relatively blue-green pigmentation. These changes frequently occur under tropical or arid conditions and may increase the difficulty of confirming some foliage characters during roguing or selection.

Pollination and isolation

The agricultural cole crops are all cross-pollinated and the different crop types are very capable of cross-pollinating. Pollination is largely achieved by insects. Most seed quality control authorities consider it important to have a greater recommended distance (up to 1500 m) between different types of *B. oleracea*, e.g. cabbages and kohlrabi, than between different cultivars of the same type, e.g. two cabbage cultivars (up to 1000 m). There are zoning schemes for the *Brassica* spp. in some countries which take the possibility of cross-pollination of the different cross-compatible types into account. In addition it is important that cole seed crops are isolated from ware crops in order to reduce the risk of those air-borne pathogens which are also subsequently seed-borne. Some national seed authorities, for example the UK and The Netherlands, have zoning schemes which include the seed crops of *Brassica* spp. (see Chapter 3).

The following isolation distances in metres are recommended for each of the cole crops included in this volume:

	Basic seed	Certified seed
Fodder kale	400	200
Cabbage	1000	500
Kohlrabi	1000	500

Pollen beetle (*Meligethes aeneus*) should be monitored from the start of anthesis; when their population reaches approximately 20 beetles per plant a spray programme should be implemented. When the population of seed weevil reaches two weevils per plant the same spray application can be applied. However, every care must be taken with both time of spray application and pollinating insect activity. Some weeds, notably *Sinapis arvensis* (charlock), are alternative hosts to both of these pests.

Seed crop establishment and management

Seed is sown at the rate of 1.5–3 kg/ha in rows 35–70 cm apart depending on the cultivar's vigour and plant size at anthesis.

Kales are biennials in terms of seed production and are sown around midsummer and are vernalized in the field during the following winter. A low level of nitrogen can be applied in the spring if the crop is slow to respond to the onset of warmer conditions.

Cabbage plants are normally raised in seedbeds and planted out or drilled direct. The sowing dates depend on the local custom and experience with specific crops. Sowing is timed so that the cultivar receives sufficient vernalization (if required), and to allow for the plants to survive hard weather but to reach anthesis and subsequently produce seed in suitable weather conditions. Sowing 60 g of seed thinly in drills 20 cm apart to produce approximately 10 seedlings per 30 cm will produce approximately 10,000 plants, subject to satisfactory germination and seedling emergence.

Young plants from the seedbeds are planted out when they have reached the five to seven leaf stage. Obvious off-types, blind plants and any showing signs of serious pathogens are discarded at this stage.

The final spacing depends on the crop and vigour of the cultivar, but a useful criterion is to ensure that the plant density is as high as possible, but there should be sufficient space to allow inspectors to examine the crop. Rows are usually 60 cm apart, although some of the smaller cabbage cultivars are planted in rows 40 cm apart. The distance between the plants within the rows is usually 30–60 cm. In some seed-production areas a 'bed' or 'set' system of up to four rows is used with a wider space between the beds; this assists subsequent roguing and/or seed crop inspections.

Some seed producers earth-up cabbage plants after they are established in their final stations. This is done as part of the cultivation programme to control weeds but it also reduces the risk of plants toppling over when they become heavier, although this is not done in high density crops.

When seed is to be produced from hearted cabbage it is usually necessary to incise the mature heads after checking trueness to type. There are several ways of achieving this but all have the objective of allowing the flower stalk to emerge unhindered by the mechanical barrier of the tightly folded leaves which encase it. A quick method is to cut the top of the head in the form of a cross, but the growing point must not be damaged.

Kohlrabi is usually direct drilled for seed production as a 'seed to seed' system, although transplanting is used for stock seed production. Unlike the other cole crops discussed above it is possible to vernalize the seed of some quick-bolting cultivars of kohlrabi. The treatment is given before a spring sowing: the seeds are pre-soaked in water for approximately 8h, then spread on moist sheets of filter paper at 20–22°C. Between 70 and 90% germination occurs in 24h at these temperatures. The moist seeds are stored at –1°C for 35–50 days (Nieuwhof, 1969). After thawing slowly the seeds are sown direct into the field. The use of this technique does not allow for roguing against early bolters and it should not be used for production of stock seed.

Overwintered crops are either left in the ground, protected from frost by covering with soil or other suitable material; otherwise they are lifted and stored in frost-free conditions. The stored roots are re-planted in early spring.

Hybrid seed production of *Brassica* species

Hybrid cultivars have become important in cabbage, kale and kohlrabi. Sporophytic self-incompatibility has been utilized for the production of F1 hybrid cultivars in all types of *Brassica oleracea* (Riggs, 1987). The ratio of female to male parent is usually 1:1 or 2:1 (Takahashi, 1987). The recommendations of maintenance breeders must be strictly adhered to when producing seed of hybrid cultivars. There are triple-cross hybrid kales.

The incidence of sibs in the production of F1 hybrid seed sometimes presents problems; Fitzgerald *et al.* (1997) have described a technique for determining sib proportion and aberrant characterization in hybrid seed using image analysis.

Roguing stages

Kale seed crops are normally only rogued once, at the start of anthesis, because unlike the vegetable brassicas which are also used as stockfeed it is not possible to view the mother plants individually.

The following characters are checked:

1. Leaf: relative length and width, degree of anthocyanin pigmentation, degree of cuticle waxiness. Serration of leaf margins.
2. Petiole: relative length, degree of anthocyanin pigmentation.
3. Main stem: shape, amount of secondary and tertiary branching, stem colour, overall plant height.
4. Characters of inflorescence: flower size and colour density.

Fodder kale: the main roguing for trueness to type is done at the start of anthesis when flower colour is confirmed. Checks on general morphology can be made earlier, although in practice this is difficult in the dense or high plant populations used for seed production.

Cabbage roguing is carried out:

1. At planting out or before heading: check for general foliage characters.
2. When heads have formed on most plants in the crop: check head characters, including shape, relative size and firmness. Discard plants that are either too early or too late for the cultivar.

Kohlrabi roguing is carried out:

1. During the final thinning: remove any plants that are bolting early, also plants that are not true to type for foliage characters, vigour or pigmentation.
2. At normal harvesting maturity of kohlrabi ('bulb'): remove early-bolting plants, check that the kohlrabi ('bulb') foliage characters and pigmentation are true to type.

Harvesting

All of the cole crops have a strong tendency to seedpod shattering. It is therefore important that appropriate action is taken to secure the potential seed crop before any is lost. Work by Still and Bradford (1998) has indicated that population-based hydrotime and ABA-time models could be used to assess physiological maturity in some *B. oleracea* types.

As the seeds ripen, the plants start to dry out and the orange-brown pigmentation of the plants is the best sign of this. The approaching maturity of the majority of seeds on a plant can also be confirmed by opening a sample of the oldest pods and therefore most mature pods, which will be the first to have become brown. The ripening seeds will also be relatively firm in response to pressing with finger and thumb.

Many seed producers prefer to cut the ripening material by hand and place in windrows or on sheets to continue drying before extracting the seed with stationary threshers. Direct combining can only be done in dry conditions and care must be taken to minimize the loss from shattering. In temperate areas with unreliable weather the ripening plants can be pulled up from the field and dried in an airy structure.

Drying

The seed should be further dried as soon as possible following harvesting. As a rule of thumb, if the seed lot's moisture is above 18% the drying air temperature should not exceed 27°C, when it is below 18% moisture content the air-drying temperature should not exceed 38°C. Whatever drying system is used, the seed depth should not impede the airflow.

Seed processing

The seeds of *Brassica* spp. crack very easily and it is therefore important that a relatively slow cylinder speed, not normally exceeding 700 rpm, is used although increased speeds may be required if the material is not very brittle. Split seeds can be separated out from a seed lot by a spiral separator. Any light or shrivelled seeds are removed by a gravity separator.

Pests and pathogens

The main seed-borne pathogens of *Brassica* species are listed in Table 13.1.

Brassica napus L. var. *napobrassica* (L.) Rchb. (Swede, Rutabaga) and *Brassica rapa* L. (including *B. campestris* L.) (Turnip)

Origins and types

The history and development of swedes and rapes and their relationships have been described by McNaughton (1984a), while the history and development of turnips and their relatives have been described by McNaughton (1984b).

Detailed Test Guideline 089 for swede and rutabaga and Test Guideline 037 for turnip are obtainable from UPOV (see Appendix).

Site

Swedes and turnips require a soil pH of 5.5–6.8; soils which have a pH below the lower end of this range should receive adequate amounts of liming materials during their preparation prior to sowing.

The general N:P:K fertilizer requirements applied during seedbed preparation are in the ratio of 1:2:2 or 1:1:1; the lower nitrogen ratio is used for turnips unless the soil's nitrogen status is low. Both crops are susceptible to boron deficiency, which causes 'brown heart' of their roots. Boronated fertilizers are used where the soils are known to be relatively low in this micronutrient. Supplementary nitrogen is usually applied as a top dressing in the spring, especially in areas or seasons in which winter rainfall will have resulted in a high rate of leaching.

However, supplementary nitrogen dressings must be carefully monitored as excessive nitrogenous fertilizers increase the incidence of lodging when the seed crop is maturing.

Seed crop establishment

Although 'seed to seed' and 'root to seed' systems are used for both crops, a 'root to seed' system is generally used for basic seed production.

When seed is direct drilled, 2–4 kg/ha are sown in rows 50–90 cm apart. The crop is thinned to approximately 4–5 cm between plants within the rows. When a crop is to be transplanted, a sowing rate of 3–4 kg/ha will provide sufficient transplants ('stecklings') for 6–10 ha. In areas that are subject to severe winter frosts the overwintering plants are protected with straw or other suitable materials. Both crop species are sown in late summer for seed production by the 'seed to seed' system, but turnips are generally less hardy than swedes and early spring sowings are made in areas where the crop would not be expected to survive the winter. For hybrid seed production of turnip the ratio of female to male parent is usually 1:1 (Takahashi, 1987).

Topping

Both crop species are 'topped' to encourage the development and growth of secondary inflorescences from the main flowering shoots. This 'topping' also reduces the overall height of the crop and the possibility of lodging at a later stage. The removal of the growing points also reduces the seed crop's range of maturity period. The top 10 cm of the terminal shoots are removed when the flowering shoots are between 30 and 40 cm high.

Pollination and isolation

Swedes and turnips are both biennials and the overwintering plants are vernalized. There is a wide range of vernalization requirements among turnip cultivars and many of those cultivars that are of agricultural importance will flower in their first year, especially from a spring sowing.

Both crop species are mainly pollinated by insects, although some wind-pollination also occurs. The vegetable cultivars of swede readily cross-pollinate with cultivars of agricultural swede, swede oil rape, swede fodder rape, kale rape and the turnip group. The agricultural turnip cultivars cross-pollinate with vegetable turnips, turnip oil rape, turnip fodder rape and the swede types listed above. The recommended isolation distances for swede or turnip is 1000 m.

Roguing

Roguing stages for swede and turnip are dependent on the system.

The seed-to-seed system

1. Early vegetative stage, before swelling of the roots: check leaf type, colour and relative height.
2. Start of anthesis: check that the flower colour and size is according to type.

The root-to-seed system

1. Early vegetative stage, before swelling of the roots: check leaf type, colour and relative height.
2. When roots are lifted for storing (or replanting, depending on local winter climate and custom): check that root shape, relative size, colour or pigmentation of root and shoulder are according to type.
3. When the replanted crop reaches anthesis: check that flower colour and flower size are according to type.

Basic seed production

Only the root to seed system is used for the production of basic seed. Roots for selection should be grown to normal market maturity by the autumn of their first year. They are then selected on root characters as described above. The early-maturing turnip cultivars are sown later than for the early market crop, otherwise the roots are too large by the late summer and autumn and have a tendency to rot during storage.

The selected roots are stored in frost-proof conditions in moist peat or sand and replanted in the late winter when no further severe frosts are expected. There is an increased tendency to replant the selected roots in polythene tunnels at the end of the winter.

Harvesting

Seed crops of both swede and turnip have a tendency to shatter readily and therefore careful cutting is required to minimize unnecessary seed loss during harvesting. Generally the best indication that the majority are near maturity is that the plant haulm turns from green-brown to parchment colour. The crop is cut with a swather and left in windrows until the seeds are mature and separate easily from their pods. The material is then either picked up from the windrows by combine or fed into a thresher. Material produced on a small scale or within polythene tunnels is hand harvested.

Drying and seed processing

This is as described above for brassicas.

Pests and pathogens

The main seed-borne pathogens of *Brassica* species are listed in Table 13.1.

Raphanus sativus L. var. *oleifera* (Fodder Radish)

Origins and types

The centre of origin of the radish is around the south Mediterranean, especially Egypt. The background and history of cultivated *Raphanus sativus* has been described by Banga (1984).

This crop is not frost hardy, therefore seed crops are sown in the spring and seed produced during the summer/autumn in the same calendar year. The duration from sowing to anthesis is relatively short in some fodder crop cultivars.

The main criteria for describing fodder radish cultivars are:

- ploidy, time of flowering and colour of petals at flowering.

A detailed Test Guideline 178 for fodder radish is obtainable from UPOV (see Appendix).

Site

Sites intended for radish seed production should not have produced a radish seed crop

Table 13.1. The main seed-borne pathogens of *Brassica* species; these pathogens may also be transmitted to the crop by other vectors.

Pathogen	Common name
Albugo candida	White blister
Alternaria brassicae	Grey leaf spot
Alternaria brassicicola	Black spot, dark leaf spot, wire stem
Leptosphaeria maculans	Dry rot, blackleg, black rot
Mycosphaerella brassicicola	Black ring spot
Phoma lingam	Dry rot, canker
Plasmodiophora brassicae	Club root, finger and toe
Rhizoctonia solani	Damping off, wire stem
Sclerotinia sclerotiorum	Watery soft rot, drop, white blight
Pseudomonas syringae pv. *maculicola*	Bacterial leaf spot
Xanthomonas campestris pv. *campestris*	Black rot
TYMV	Turnip yellow mosaic virus

during the previous 5 years. The wild radish *Raphanus raphanistrum* is a contaminant of major importance from the point of view of cross-pollination and also as an undesirable weed seed in the harvested seed crop from which it cannot be separated by seed processing.

Pollination and isolation

Fodder radish is cross-pollinated and isolation distances between any radish seed crop should be at least 1000 m. *Raphanus sativus* is not cross-compatible with *Brassica* spp.

Crop establishment

The radish crop requires a soil with a pH of 5.5–6.8. It is advantageous to incorporate a nitrogenous fertilizer at the rate of 100 kg/ha during site preparation. The seed crop is sown at the rate of 6 kg/ha in rows 45 cm apart. Larger seed crop areas of approximately 4 ha or more are preferable in order to minimize loss from feeding birds during ripening.

Roguing

This is done at the start of anthesis. The main character examined during roguing is the flower colour, to confirm that it corresponds to that of the cultivar's description. During the same operation, any other off-types that display morphological characters which are not in accord with the cultivar's description are removed from the seed crop. The final removal of any wild radish plants should also be done during this operation.

Harvesting

The ripening seed crop is very vulnerable to birds, therefore plans and arrangements should be made in advance for appropriate deterrents.

The 'seed pods' (silicula) are rather fleshy and slow to ripen. The pointed seed pods start to turn a reddish-brown colour. The best indication for cutting the crop and drying in windrows is when seeds in the 'pods' start to become brown. The cut crop can take up to 3 weeks for further drying in temperate environments. The windrowed crop is finally picked up and threshed by combine.

Drying

The threshed seed should be immediately dried to 10% moisture content. Once the seed processing has been completed the seed is further dried to a moisture content of 5% for vapour-proof storage.

Seed processing

The harvested seed lot can be upgraded by an air/screen cleaner.

Pests and pathogens

The main seed-borne pathogens of fodder radish are listed in Table 13.2.

Table 13.2. The main seed-borne pathogens of fodder radish; these pathogens may also be transmitted to the crop by other vectors.

Pathogen	Common names
Alternaria alternata	
Alternaria brassicae	Grey leaf spot
Alternaria brassicicola	Black leaf spot
Alternaria raphani syn. *A. matthiolae*	Leaf spot
Colletotrichum higginsianum	Anthracnose, leaf spot
Gibberella avenacea	Root and stem rot
Leptosphaeria maculans	Black leg
Rhizoctonia solani	Damping-off, canker
Xanthomonas campestris pv. *raphani*	Bacterial spot
RYEV	Radish yellow edge virus
TSV	Tobacco streak virus
TYMV	Turnip mosaic virus

Chenopodiaceae – Fodder Beet and Mangolds

Beta vulgaris L. subsp. *vulgaris* (Fodder Beet, Mangel, Mangold, Mangelwurzel)

Origin and types

The subspecies *Beta vulgaris* ssp. *maritima* is generally believed to be the original wild type from which the above domesticated *B. vulgaris* subspecies and morphological types were derived; these have been further selected and developed over many years of cultivation. This wild species can be a problem weed in the cultivated forms from the point of view of pollen contamination and also admixture of its highly undesirable seeds in harvested seed lots from which it cannot be separated by processing.

Other types include sugarbeet, which is an industrial crop cultivated for sugar production and is not discussed in this volume (see Kelly and George, 1998c); the other crop is red beet, also known as garden beet, the seed production details of which have been discussed by George (2009).

A detailed Test Guideline 150 for fodder beet is obtainable from UPOV (see Appendix).

Fodder beet

The roots have a high dry-matter content; traditionally these have produced longer and narrower roots than mangels and are deeper in the soil, which can present difficulties when lifting. However, the fodder beets have been significantly improved by plant breeders and there are now cultivars with improved root shape, which are easier to lift and also have a high dry-matter content. Many of the improvements in fodder beet cultivars have come about as a result of sugarbeet development and breeding. The improved cultivars are based on crosses between sugarbeet and mangel.

Mangel

The roots have a relatively low dry-matter content, and there is usually a clearly defined morphology of root shape, which sits 'prouder' out of the soil. This crop has lost favour as a result of plant-breeding improvements of fodder beet; however, it is still popular in those agricultural systems in which the crop has to be lifted by hand.

Seed production of fodder beet

Site

There should be a 5-year break between *Beta* seed crops, with the additional proviso that all beet crops should be excluded for the 2 years before the *Beta* seed crop.

Every care should be taken to eradicate any possibility of dormant seed and growing plants of the weed beet on the site. The so called 'weed beet' are annual forms of beet and may be wild beet or crosses between wild beet and fodder beet, or they may have arisen from bolters in a normal beet crop. Weed beet sheds seed on arable land and can survive and multiply over many years unless efficiently controlled. These sources of contamination also include neighbouring fields of other crop species within the specified seed class isolation requirements. Weed beet seeds cannot be removed from seed lots of fodder beet seed during seed processing. The other weed seeds that are difficult to remove from a fodder beet or mangel seed lot during processing include:

- *Atriplex patula* – common orache;
- *Chenopodium album* – fat hen;
- *Galium aparine* – cleavers;
- *Polygonum* spp. – bindweed; and
- *Stellaria media* – common chickweed.

Isolation and pollination

All the subspecies of *B. vulgaris* are wind-pollinated and also cross-compatible with one another; there can also be cross-pollination by insects. Adequate isolation distances are therefore essential between each and all of the subspecies and their cultivars too, according to the seed class being produced.

The subspecies include fodder beet and mangels, sugarbeet, garden (red) beet, wild (sea) beet, spinach beet and sea-kale beet (Swiss chard). The OECD (1995) specifies the following distances:

1. No isolation distance is necessary between seed crops using the same pollinator.
2. All seed crops to produce Basic Seed from any pollen source of the genus *Beta*: 1000 m.
3. All seed crops to produce Certified Seed of fodder beet:

(i) from any pollen source of the genus *Beta* not included below: 1000 m;
(ii) the intended pollinator or one of the pollinators being diploid, from tetraploid fodder beet pollen sources: 600 m;
(iii) the intended pollinator being exclusively tetraploid, from tetraploid fodder beet pollen sources: 600 m;
(iv) from fodder beet sources the ploidy of which is unknown: 600 m;
(v) the intended pollinator, or one of the pollinators, being diploid fodder beet sources: 300 m;
(vi) the intended pollinator being exclusively tetraploid, from tetraploid fodder beet pollen sources: 300 m; and
(vii) between two seed production fields in which male sterility is not used: 300 m.

There are zoning schemes in some areas, which restrict specific cultivars in specified zones. These schemes may be voluntary agreements between seed growers and plant breeders, or the restrictions may have national or regional legal status.

Additional measures to maintain or improve cultivar purity: discard strip techniques

Isolation experiments were done by Dark (1971), working with sugarbeet. He subsequently made the following recommendations:

1. Breeders' seed crops should be produced in protected structures.
2. Basic seed crops should be isolated by 1000 m, and that a 4.6 m strip on the outside of the crop should be discarded. However, in practice the seed from the outside strip can be placed in a lower seed category when no discrepancies of isolation requirements have been reported.
3. All seed crops should be as nearly square as possible so as to present the shortest possible proportion of outside edge.

Seed crop establishment

The crop species is a biennial that requires vernalization at the appropriate development stage followed by longer day length to induce flowering. There are three possible seed-production systems; these are listed below along with other points with regard to their suitability for specific circumstances.

1. Mature roots are selected in the field seeded *in situ*. The non-selected roots are either destroyed, or ideally taken and used as a ware crop. Alternatively, the selected roots are lifted and moved to an isolation cage or structure for seeding. This method was originally used for mangels when roots were selected on morphological characters (root shape and outer colour).
2. Seed is sown in a seedbed for the production of young plants, usually referred to as 'stecklings'. The stecklings are either transplanted to their final seeding site in the autumn or at the end of winter. This is now mainly used during breeding programmes. It is also used where there is a large labour force.
3. Seed is sown and the plants are allowed to mature and produce seed in situ. This system is largely used for the commercial production of fodder beet seed. The seed is either sown on bare soil in midsummer, or earlier under the cover of a suitable cover crop; a cereal is normally used in this method. The cover crop should not be too dense or the beet seedlings will be suppressed. The optimum plant population is 300,000/ha. When sowing on open ground the row spacing is 19 cm apart with plants 25 cm apart within the rows; the reason for this is to achieve the optimum spacing. If the crop is undersown the rows can be closer together and the plants spaced wider apart within the rows. If herbicides are to be used care should be taken to ensure that they are suitable for both crop species. It may be necessary to

cultivate inter-row after the cover crop has been harvested.

Roguing

Current roguing policy concentrates on removing early bolters until the majority of the seed crop reaches anthesis. The other main concern is the presence of mother plants and/or weeds showing disease symptoms, especially symptoms of virus yellows and any identifiable seed-borne pathogens. These should be checked from the steckling bed stage onwards.

Additional attention should be given to any plant from the seedling or steckling stage onwards which show anthocyanin (red) pigmentation; these off-types should be removed when identified as they indicate previous generation contamination by red beet.

The surrounding area, including waste plots and private gardens, should also be checked for early bolters and wild beet.

Harvesting

Beet seeds do not ripen evenly on the mother plant and individual seeds shed easily when ripe. The maturity and ripeness of the first seeds is indicated when the plants start to turn yellow. Some growers use a desiccant when the first seeds start to ripen, the advantages being that there is less non-seed plant material to handle and there is a reduction of loss from shattering; as an alternative, the desiccant can be applied to the windrowed crop. Furthermore, beet 'seeds' are not normally adversely affected by desiccants. The crop should be cut at this stage, leaving high stubble and windrowed; smaller areas can be cut by hand and tied in bundles and stooked. The higher stubble assists with aeration of the windrowed crop and keeps the material off the ground, which is especially important in wet conditions. From this stage onwards it is best to minimize the movement of the cut material. Ideally for stooked crops they should be combined or threshed by a mobile machine moving from stook to stook. Windrowed crops can be picked up by combine.

Drying

Seed lots with their moisture content greater than 10% should be dried down to 8% immediately following threshing and seed processing prior to storage. The drying can be done by either batch or continuous-flow driers; the drying air temperature should not exceed 38°C.

Seed processing

Multigerm seed is first rubbed to separate into single seeds. The standards given by OECD (1995) for precision seed of fodder beet are:

> In seed of cultivars with more than 85% diploids, at least 58% of the germinated clusters shall give single seedlings; in other seed at least 63% of the germinated clusters shall give single seedlings; in both no more than 5% shall give three or more seedlings.

Pests and pathogens

The main seed-borne pests and pathogens of *B. vulgaris* spp. are listed in Table 13.3.

Chenopodium quinoa Willd. (Quinoa)

Origins and types

Quinoa was domesticated in the high-elevation Andean complex probably some 8000 to 4000 years ago. The crop was a staple food of the Incas and has become an important domesticated grain crop in the mid- to higher altitudes of the Andes where it is hardy. The seed crop is regarded as a very good source of protein. This species has created interest for its production in other high altitude areas such as the north of Pakistan and north-eastern India. Risi and Galway (1994) have discussed the crop's possible role in current agriculture. The cultivated quinoa species usually produce a satisfactory crop in some areas where other cereals cannot produce an acceptable crop.

There is a lesser known quinoa species *Chenopodium pallidicaule*, which is cultivated

Table 13.3. The main seed-borne pathogens and pests of *Beta vulgaris* species; these may also be transmitted to the crop by other vectors.

Pest or pathogen	Common names
Alternaria alternata	Seedling rot, leaf spot
Cercospora beticola	Leaf spot (of warmer climates)
Colletotrichum dematium f. *spinaciae*	Anthracnose
Erysiphe betae	Powdery mildew
Fusarium spp.	Fusarium yellows, fusarium wilt
Peronospora farinosa f.sp. *betae*	Downy mildew
Pleospora betae	Blackleg, damping-off, leaf spot
Ramularia beticola	Leaf spot (of cooler climates)
Pseudomonas syringae pv. *aptata*	Bacterial blight, leaf spot, black streak, black spot
AMV	Arabis mosaic virus
TBRV	Tomato black ring virus
RpRSV	Raspberry ringspot virus
LRSV	Lychnis ringspot virus
Ditylenchus dipsaci	Eelworm canker

in the higher altitudes of Bolivia and Peru; this species is more tolerant to lower temperatures, drought and soil salinity than *C. quinoa*. A third species is *Chenopodium nuttaliae*, commonly known as huauzontle, which is cultivated in Mexico. However, this is grown and prepared as a vegetable for its broccoli-like heads of compact green flower buds.

Five types of *C. quinoa* have been described by Fleming and Galwey (1995), although the authors found that there is significant overlap between the types described in their classification.

1. *Valley type*: 2000–4000 m; branched, growing 2–3 m tall; lax inflorescences; late maturing; very diverse.
2. *Altiplano type*: 4000 m; shorter and earlier than the valley type; branched or unbranched; mostly pigmented; small inflorescences.
3. *Solar type*: 4000 m; soils above pH 8; red pigmented plants; black seeds.
4. *Sea-level type*: earliest to mature; short, unbranched, green plants; small, yellow translucent seeds.
5. *Subtropical type*: only one plant has been found in Bolivia; intense green-orange at maturity; small yellow-orange seeds.

In addition to the above descriptions by Fleming and Galwey (1995) there are day-neutral and long-day cultivars and also saponin-free or saponin-present cultivars. The possibility of electrophoretic characterization of quinoa seed proteins can be used for distinguishing between cultivars (Fairbanks *et al.*, 1990). Ripened seed of this species shatters very readily; however, plant breeders have produced lines and cultivars that are less prone to shattering.

Site

Quinoa is frequently cultivated in a rotation that includes paddy rice, pulses or other arable crops depending on the locally preferred crops. Quinoa seed can survive in the soil for several years, therefore there should be a rotation of at least 3 years prior to production of a seed crop; a longer break period should be planned for high seed classes. Quinoa seed is very small and it is therefore virtually impossible to separate from other small-seeded weed species in the seed lot. The important weed species include indigenous and wild quinoa species, including *C. quinoa* var. *melanospermum*.

Pollination and isolation

The species is generally regarded as self-fertilized although Simmonds (1984) reported up to 9% of cross-pollination. The plants are gynomonoecious (i.e. there are both hermaphrodite and female flowers on the same plant).

Seed crop establishment

The stock seed is sown in moist soil at the rate of 12–20 kg/ha in rows 20–50 cm apart. Sowing or drilling depth should not exceed 1–2 cm. Seedlings usually emerge within 7 days from sowing. The crop responds to nitrogen, which can either be incorporated as a base dressing or top dressing after emergence, depending on the available nitrogen and potential leaching rate. Some growers who adopt the wider row spacing earth the young plants up following seedling establishment. The crop is cultivated to remove weeds before their competition is likely to interfere with the establishment of the quinoa seed crop.

Roguing

The roguing criteria include type of inflorescence. Individual types have an associated seed colour although this is not of use when roguing although it can on occasions indicate the cultivar purity of a given seed lot.

Harvesting

The ripe seeds are almost hard when ripe and difficult to indent with a thumb nail. Large production areas can be combined direct when the crop plants are completely dry; otherwise the ripe crop can be harvested by hand and then threshed by traditional hand methods or with a small thresher.

Drying

No information on the safe procedures and temperatures for drying quinoa seed are available.

Pests and pathogens

There are no pests or pathogen species of quinoa recorded as proven to be seed-borne.

Further Reading

Cruciferae

Bowring, J.D.C. and Day, M.J. (1977) Variety maintenance for swedes and kale. *Journal of the National Institute of Agricultural Botany* 14, 312–320.

Beet

Dark, S.O.S. (1971) Experiments on the cross-pollination of sugar beet in the field. *Journal of the National Institute of Agricultural Botany* 12, 242–266.

Quinoa

Fairbanks, D.J., Burgener, K., Robinson, L.R., Anderson, W.R. and Ballon, E. (1990) Electrophoretic characterisation of quinoa seed proteins. *Plant Breeding* 104, 190–195.

Fleming, J.E. and Galwey, N.W. (1995) Quinoa (*Chenopodium quinoa*). In: Williams, J.T. (ed.) *Cereals and Pseudo Cereals*. Chapman and Hall, New York, pp.3–84.

14

Amaranthaceae

The Grain Amaranths

Origins and types

The genus *Amaranthus* has cultivated species suitable for grain production and protein-rich leaves which are used as pot herbs or green vegetables. Although originally domesticated by the early civilizations in tropical and subtropical America, the grain amaranths are currently more important in the Indian subcontinent. The history of the grain amaranths has been described by Sauer (1976). The *Amaranthus* spp. are herbaceous annuals; the main species cultivated for grain production are:

- *Amaranthus caudatus* L., syn. *A. edulis* Speg. – originates from South America and is cultivated in the Andean regions of Argentina, Bolivia and Peru. This species, which has several common names, including kiwicha, quinoa and Inca wheat, can be cultivated for grain and/or as a green leaf vegetable and the genus is regarded as a dual purpose crop. The nutritional value of the grain is reputed to be as good as milk. Seed production of the vegetable amaranths is described in George (2009).
- *Amaranthus caudatus* ssp. *mantegazzianus* Passer – this species also originates from South America, and is cultivated predominantly for its grain.
- *Amaranthus hypochondriacus* L. syn. *A. leucocarpus* S. Wats. – originates from Mexico and parts of Central America, and is widely cultivated for its grain.

There has been an increasing interest in producing commercial grain crops from the amaranths in areas far removed from their origins, for example Austria and North America. A review of the grain amaranths has been made by Williams and Brenner (1995).

Classification of cultivars

Detailed Test Guideline 247 for Grain Amaranth is obtainable from UPOV (see Appendix).

Vernon (1999) has described some of the *Amaranthus* cultivars that have been derived from material originating in Central and South America. The grain amaranths produce a substantial sized plant, usually with strong apical dominance (Fig. 14.1).

Site

The site for seed production should not have grown amaranths in the previous year; it should be free of *Amaranthus spinosa*, the weed amaranth, which is also known as prickly calalue or thorny amaranth. Ideally the black-seeded

Fig. 14.1. A grain amaranth plant bearing a good potential seed crop.

types of amaranth should be excluded from the grain crop, but this generally requires a high degree of purity in the stock seed. The soil pH should be approximately 6.5; lime-induced chlorosis can occur on soils with a pH above 6.5. The crop responds positively to a high soil potassium level. Nitrogen should be sparingly applied as the taller types for grain production lodge very readily.

Pollination and isolation

The amaranths are largely self-fertilized although up to 30% cross-fertilization can take place from wind-dispersed pollen. The recommended isolation distance is 400 m for Foundation Seed while 200 m is recommended for the production of Certified Seed although specific requirements differ from one country to another. Greater distances should be allowed between grain amaranths and the vegetable types, especially from the vegetable types that have red or patterned leaf pigmentation.

Seed crop establishment

The seed crop can be established either by transplanting bed-raised seedlings or direct sowing. Assuming a weed-free seedbed, seed is sown at a rate to obtain a stand of approximately 360,000 plants/ha, in rows 45–60 cm apart with an optimum spacing within the rows of 8 cm. These high plant populations tend to increase apical dominance and make the crop more suitable for a once-over harvest, whether by hand or mechanization. The sowing rate in seedbeds is 2 g/m^2, which would be expected to provide approximately 1000 transplants 3 weeks after sowing. Seedlings are transplanted at the same final plant density as a drilled crop.

Roguing

The seed crop should be rogued at the following stages:

1. At planting out or thinning young plants: check that plants are true to type, and that early flowering plants are removed; appropriate vegetative characters, including leaf number per plant, leaf shape and size according to plant stage are confirmed; the degree and pattern of leaf and stem pigmentation is confirmed according to the cultivar.
2. Before the start of anthesis: as above; in addition check that the plants are of appropriate height and that the degree of branching is according to the cultivar. Check that isolation is satisfactory for the seed class.
3. At the start of anthesis: confirm appropriate plant height, time of flowering, inflorescence and individual flower colour; also general morphology of the plants.
4. A final check should be made to rogue out plants producing black coated seeds as soon as they can be identified and before they contribute more to the crop's gene pool. This is especially important for those seed classes to be used for further multiplication.

Harvesting

Small-scale plots are usually hand harvested by making up to three passes through the crop, each time cutting predominantly ripe terminal and lateral seed heads. The cut heads are further dried. Large-scale plots, especially of low-shedding cultivars, can be combined. Because amaranth seed is relatively small, elevator casings and small gaps in apparatus should be sealed to minimize losses.

Drying

The harvested seed should be immediately dried down to below 10% if necessary.

Seed processing

Small or large seed lots can be cleaned by an appropriate sized air/screen cleaner. As with the machines used for harvesting this seed crop, gaps should be sealed to prevent seed loss during processing.

Pests and pathogens

The two main seed-borne pathogens of amaranthus are:

- *Alternaria amaranthi* (Peck) Van Hook – blight, stem rot;
- SLRSV – strawberry latent ring spot virus.

These two pathogens may also be transmitted by other vectors.

Further Reading

George, R.A.T. (2009) *Vegetable Seed Production*, 3rd edn. CAB International, Wallingford, UK.

Williams, J.T. and Brenner, D. (1995) Grain amaranth (*Amaranthus* spp.). In: Williams, J.T. (ed.) *Cereals and Pseudocereals.* Chapman and Hall, New York, pp.129–186.

Appendix

UPOV Test Guidelines that Relate to the Agricultural Crops Included in this Volume

All Test Guidelines should be read in conjunction with document TG/1/3 'General Introduction to the Examination of Distinctness, Uniformity and Stability and the Development of Harmonized Descriptions of New Varieties of Plants', which replaces document TG/1/2.

Further information with regard to obtaining individual crop Test Guidelines can be obtained from UPOV at: www.upov.int/en/publications/tg_rom/tg_index.html (Accessed 20 February 2011).

Number	Name
019	Barley
206	Broad bean
180	*Bromus auleticus*
048	Cabbage
143	Chick-pea
031	Cocksfoot
150	Fodder beet
178	Fodder radish
012	French bean
247	Grain amaranth
093	Groundnut
065	Kohlrabi
210	Lentil
006	Lucerne
002	Maize
020	Oats
007	Pea
260	Pearl millet
036	Rape seed
005	Red clover
180	Rescue grass, Alaska brome-grass, *Bromus auleticus*

Continued

Continued

Number	Name
016	Rice
058	Rye
004	Ryegrass
134	Safflower
122	Sorghum
170	Subterranean clover
081	Sunflower
089	Swede, Rutabaga
034	Timothy
121	Triticale
037	Turnip
085	Turnip rape
003	Wheat
038	White clover

References

Anonymous (1952) *Proceedings VI International Grassland Congress*, Pennsylvania, USA, p. 855.

Anonymous (1980) *International Code for the Nomenclature of Cultivated Plants.* The International Bureau for Plant Taxonomy and Nomenclature, Utrecht, Netherlands, pp.12–13.

Anonymous (1994) *Genebank Standards.* Food and Agriculture Organization of the United Nations, Rome; International Board for Plant Genetic Resources, Rome.

Anonymous (1995) Getting Farmers Involved in Research. *CGIAR News* 2(2), 9–10.

AOSA (1983) *Seed Vigour Testing Handbook.* Contribution No. 32 to the Handbook of Seed Testing, Association of Official Seed Analysts, Ithaka, New York.

AOSA (1993) Association of Official Seed Analysts, Inc. Bylaws. *Newsletter of the Association of Official Seed Analysts* 63(3), 93–105.

AOSA (2008) *Rules for Seed Testing.* AOSA, Ithaka, New York.

Atkins, E.L., Anderson, L.D., Kellum, D. and Neuman, K.W. (1977) *Protecting Honey Bees from Pesticides.* Leaflet 2883, Division of Agricultural Sciences, University of California, Berkeley, California.

Austin, R.B. (1972) Effects of environment before harvesting on viability. In: Roberts, E.H. (ed.) *Viability of Seeds.* Chapman and Hall, London and Syracuse University Press, Syracuse, New York, pp.114–149.

Banga, O. (1984) Radish. In: Simmonds, N.W. (ed.) *Evolution of Crop Plants.* Longman, London and New York, pp.60–62.

Bateman, A.J. (1946) Genetical aspects of seed-growing. *Nature* 157, 752–755.

BCPC (2009) *Seed Production and Treatment in a Changing Environment.* Proceedings No. 83. BCPC Publications, Thornton Heath, Surrey, UK.

BCPC (2010) *The GM Crop Manual.* BCPC Publications, Thornton Heath, Surrey, UK.

Biddle, A.J. and King, J.M. (1977) Effect of harvesting on pea seed quality. *Acta Horticulturae* 83, 77–81.

Bonina, J. and Cantliffe, D.J. (2005) *Seed Production and Seed Sources of Organic Vegetables.* IFAS Extension, University of Florida, Gainesville, Florida.

Boonman, J.G. (1973) The effect of harvest date on seed yields in varieties *Setaria sphacelata, Chloris gayana* and *Panicum coloratum. Netherlands Journal of Agricultural Sciences* 21, 3–11.

Borrill, M. (1984) Temperate Grasses *Lolium, Festuca, Dactylis, Phleum, Bromus.* In: Simmonds, N.W. (ed.) *Evolution of Crop Plants.* Longman, London and New York, pp.137–142.

Bradnock, W.T. (1998a) Association of Official Seed Certifying Agencies (AOSCA). In: Kelly, A.F. and George, R.A.T. (eds) *Encyclopaedia of Seed Production of World Crops.* John Wiley and Sons, Chichester, UK, pp.28–31.

Bradnock, W.T. (1998b) Association of Official Seed Analysts (AOSA). In: Kelly, A.F. and George, R.A.T. (eds) *Encyclopaedia of Seed Production of World Crops.* John Wiley and Sons, Chichester, UK, pp.34–36.

Brouwer, W. (1949) Steigerung der Erträge der Hülsenfrüchte durch Beregnung sowie Fragen der Bodenuntersuchung und Düngung. *Zeitschrift fur Acker und Pflanzenbau* 91, 319–346.
Brouwer, W. (1959) *Die Feldberegnung*, 4th edn. DLG-Verlag, Frankfurt/Main.
Brown, J., Brown, A.P., Davis, J.B. and Erickson, D. (1997) Intergeneric hybridization between *Sinapis alba* and *Brassica napus. Euphytica* 93, 163–168.
Browning, T. and George, R.A.T. (1981a) The effects of nitrogen and phosphorus on seed yield and composition in peas. *Plant and Soil* 61, 485–488.
Browning, T. and George, R.A.T. (1981b) The effects of mother plant nitrogen and phosphorus nutrition on hollow heart and blanching of pea (*Pisum sativum* L.) seed. *Journal of Experimental Botany* 32(130), 1085–1090.
Browning, T., Gavras, M. and George, R.A.T. (1982) *Proceedings XXI International Horticultural Congress*, Vol. 2. Hamburg, Germany, p.2039.
Caetano-Anolles, G. (1996) Fingerprinting nucleic acids with arbitrary oligonucleotide primers. *Agro-Food Industry Hi-Tech*, Jan/Feb, 26–35.
Cane, J.H. (2008) A native ground nesting bee, *Nomia melanderi*, sustainably managed to pollinate alfalfa across an intensively agricultural landscape. *Apidologie* 39, 315–323.
Chang, T.T. (1984) *Oryza sativa* and *Oryza glaberrima*. In: Simmonds N.W. (ed.) *Evolution of Crop Plants*. Longman, London and New York, pp.98–104.
Christensen, C.M. (1972) Microflora and seed deterioration. In: Roberts, E.H. (ed.) *Viability of Seeds*. Chapman and Hall, London and Syracuse University Press, Syracuse, New York, pp.59–93.
Cooke, R.J. (1992) *Electrophoresis Testing*. International Seed Testing Association, Zurich, 36 pp.
Cooke, R.J. (1995a) Gel electrophoresis for identification of plant varieties. *Journal of Chromatography* 698, 281–299.
Cooke, R.J. (1995b) Varietal identification of crop plants. In: Skerritt, J. and Appels, R. (eds) *New Diagnostics in Crop Sciences*. CAB International, Wallingford, UK, pp.33–63.
Cooke, R.J. (1995c) Variety identification: modern techniques and applications. In: Basra, A.S. (ed.) *Seed Quality: Basic Mechanisms and Agricultural Implications*. Food Products Press, New York, pp. 279–318.
Cooke, R.J. and Reeves, J.C. (1998) Cultivar identification: review of new methods. In: Kelly, A.F. and George, R.A.T. (eds) *Encyclopaedia of Seed Production of World Crops*. John Wiley and Sons, Chichester, UK, pp.82–102.
Cooke, R.J., Higgins, J., Morgan, A.G. and Evans, J.L. (1985) The use of a vanillin test for the detection of tannins in cultivars of *Vicia faba* L. *Journal of the National Institute of Agricultural Botany* 17, 139–143.
Cooper, W. and MacLeod, J. (1998) Introduction to genetically modified crops. *Plant Varieties and Seeds* 11, 131–141.
Copeland, L.O. and McDonald, M.B. (1995) *Principles of Seed Science and Technology*, 3rd edn. Chapman and Hall, New York.
Cromarty, A.S., Ellis, R.H. and Roberts, E.H. (1982) *The Design of Seed Storage Facilities for Genetic Conservation*. International Board for Plant Genetic Resources, Rome.
CRS (2002) *Seed Vouchers and Fairs: a Manual for Seed-based Agricultural Recovery*. Catholic Relief Services, developed in collaboration with International Crops Research Institute for the Semi-Arid Tropics and Overseas Development Institute, Nairobi.
Dark, S.O.S. (1971) Experiments on the cross-pollination of sugar beet in the field. *Journal of the National Institute of Agricultural Botany* 12, 242–266.
Davies, D.R. (1984) Peas. In: Simmonds, N.W. (ed.) *Evolution of Crop Plants*. Longman, London and New York, pp.172–174.
Davis, J.H.C. (1997) *Phaseolus* beans. In: Wien, H.C. (ed.) *The Physiology of Vegetable Crops*. CAB International, Wallingford and New York, pp.409–428.
Dayday, H. (1955) Cyanogenesis in strains of white clover. *Journal of the British Grassland Society* 10, 266–274.
Delouche, J.C. (1980) Environmental effects on seed development and seed quality. *Hortscience* 15, 13–18.
Doggett, H. and Rao, K.E.P. (1995) Sorghum, *Sorghum bicolor* (Gramineae-Androponeau). In: Smartt, J. and Simmonds, N.W. (eds) *Evolution of Crop Plants*. Longman Scientific and Technical, Harlow, UK, pp.173–180.
Dowker, B.D., Bowman, A.R.A. and Faulkner, G.J. (1971) The effect of selection during multiplication on the bolting resistance and internal quality of Avon early red beet. *Journal of Horticultural Science* 46(3), 307–311.

Drijfhout, E. (1981) Maintenance breeding of beans. In: Feistritzer, W.P. (ed.) *Seeds*. FAO Plant Production and Protection Paper 39, FAO, Rome, pp.140–150.

Ellis, R.H. and Roberts, E.H. (1980) Improved equations for the prediction of seed longevity. *Annals of Botany* 45, 13–30.

Ellis, R.H. and Roberts, E. (1981) The quantification of ageing and survival in orthodox seeds. *Seed Science and Technology* 9, 373–409.

Ellis, R.H., Hong, T.D., Astley, D., Pinnegar, A.E. and Kraak, H.L. (1996) Survival of dry and ultra-dry seeds of carrot, groundnut, lettuce, oilseed rape and onion during five years' hermatic storage at two temperatures. *Seed Science and Technology* 24(2), 347–358.

Eponou, T. (1996) Linkages between research and technology users in Africa: the situation and how to improve it. *ISNAR Briefing Paper Number 31*, International Service for National Agricultural Research, The Hague.

Evans, G.M. (1984) Rye. In: Simmonds, N.W. (ed.) *Evolution of Crop Plants*. Longman, London and New York, pp.108–111.

Fairbanks, D.J., Burgener, K., Robinson, L.R., Anderson, W.R. and Ballon, E. (1990) Electrophoretic characterisation of quinoa seed proteins. *Plant Breeding* 104, 190–195.

FAO (1993) *Quality Declared Seed*. FAO Plant Production and Protection Paper 117, FAO, Rome.

FAO (1999) *Restoring Farmers' Seed Systems in Disaster Situations*. FAO Plant Production and Protection Paper 150, FAO, Rome.

FAO (2001a) *Incorporating Nutrition Considerations into Agricultural Research Plans and Programmes*. FAO, Rome.

FAO (2001b) *Seed Policy and Programmes for the Central and Eastern European Countries, Commonwealth of Independent States and other Countries in Transition*. FAO Plant Production and Protection Paper 168, FAO, Rome.

FAO (2004) *Towards Effective and Sustainable Seed Relief Activities*. FAO Plant Production and Protection Paper 181, FAO, Rome.

FAO (2006) *Quality Declared Seed System*. FAO Plant Production and Protection Paper 185, FAO, Rome.

Faulkner, G.J. (1983) *Maintenance, Testing and Seed Production of Vegetable Stocks*. Vegetable Research Trust, NVRS, Wellesbourne, Warwick, UK.

Feistritzer, W.P. (1981) The FAO Seed Improvement and Development Programme (SIDP). *Seed Science and Technology* 9(1), 37–45.

Feldman, M. (1984) Wheats. In: Simmonds, N.W. (ed.) *Evolution of Crop Plants*. Longman, London and New York, pp.120–128.

Fitzgerald, D.M., Barry, D., Dawson, P.R. and Cassels, A.C. (1997) The application of image analysis in determining sib proportion and aberrant characterization in F1 hybrid Brassica populations. *Seed Science and Technology* 25(3), 503–509.

Fleming, J.E. and Galwey, N.W. (1995) Quinoa (*Chenopodium quinoa*). In: Williams, J.T. (ed.) *Cereals and Pseudo Cereals*. Chapman and Hall, New York, pp.3–84.

Forsberg, G., Anderson, S. and Johnsson, L. (2000) Evaluation of hot, humid air seed treatment in thin layers and fluidized beds for seed pathogen sanitation. *Journal of Plant Disease and Protection* 109, 357–370.

Garry, G., Jeuffroy, M.H. and Tivoli, R. (1998) Effects of ascochyta blight *Mycosphaerella pinodes* (Berk. and Blox.) on biomass production, seed number and seed weight of dried-pea (*Pisum sativum* L.) as affected by plant growth stage and disease intensity. *Annals of Applied Biology* 132(1), 49–59.

Gaunt, R.E. and Liew, R.S.S. (1981) Control of diseases in New Zealand broad bean seed crops. *Acta Horticulturae* 111, 109–112.

Gavras, M. (1981) The influence of mineral nutrition, stage of harvest and flower position on seed yield and quality of *Phaseolus vulgaris* L. PhD thesis, University of Bath, Bath, UK.

George, R.A.T. (2009) *Vegetable Seed Production*, 3rd edn. CAB International, Wallingford, UK.

Golden Rice (2011) Available at: www.goldenrice (Accessed 7 February 2011).

Goulson, B., Lye, D.G. and Darvill, B. (2008) Decline and conservation of bumble bees. *Annual Review of Entomology* 53, 191–208.

Graeme, J.D., Pannel, D.J. and Revell, C.R. (2009) Economic contribution of French Seradella (*Ornithopus sativus* Brot.) pasture to integrated weed management in Western Australian mixed-farming systems: an application of compressed annealing. *Australian Journal of Agricultural and Resource Economics* 53(2), 193–212.

Gray, D. (1994) Large-scale seed priming techniques and their integration with crop protection. In: Martin, T. (ed.) *Seed Treatment: Progress and Prospects*. British Crop Protection Council, Surrey, UK, pp.353–362.

Greengrass, B. (1998) International Union for the Protection of New Varieties of Plants. In: Kelly, A.F. and George, R.A.T. (eds) *Encyclopaedia of Seed Production of World Crops*. John Wiley and Sons, Chichester, UK, pp.19–23.

Griffiths, D.J. (1950) The liability of seed crops of ryegrass (*Lolium perenne*) to contamination by wind-borne pollen. *Journal of Agricultural Science* 40, 19–38.

Groot, S.P.C., Oosterveld, O., van der Wolf J., Jalink, W.M., Langeraak, C.J. and van den Bulk, R.W. (2004) The role of ISTA and seed science in assuring organic farmers with high quality seeds. In: FAO (2004) *First World Conference on Organic Seed Proceedings*. The International Federation for Organic Movements (IFOAM), The International Seed Federation (ISF) and FAO, Rome, pp.9–12.

Hacquet, J. (1990) Genetic variability and climate factors affecting lucerne seed production. *Journal of Applied Seed Production* 8, 59–67.

Haldeman, C. (2008) ISTA and Biotech/GM crops. *Seed Testing International, ISTA News Bulletin* 136, October 2008, 3–5.

Halmer, P. (1994) The development of quality seed treatments in commercial practice. In: Martin, T.J. (ed.) *Seed Treatment: Progress and Prospects*. The British Crop Protection Council, Thornton Heath, UK, pp.363–374.

Hancock, J.F. (2004) *Plant Evolution and the Origin of Crop Species*, 2nd edn. CAB International, Wallingford, UK.

Hardegree, S.P. and Van Vactor, S.S. (2000) Germination and emergence of primed grass seeds under field and simulated field temperature regimes. *Annals of Botany* 85, 379–390.

Harlan, J.R. (1984a) Barley. In: Simmonds, N.W. (ed.) *Evolution of Crop Plants*. Longman, London and New York, pp.93–98.

Harlan, J.R. (1984b) Tropical and sub-tropical grasses. In: Simmonds, N.W. (ed.) *Evolution of Crop Plants*. Longman, London and New York, pp.142–144.

Harrington, J.F. (1963) Practical advice and instructions on seed storage. *Proceedings of The International Seed Testing Association* 28, 989–994.

Hartley, W. and Williams, G.L. (1956) Centres of distribution of cultivated pasture grasses. *Proceedings 7th International Grassland Conference*, New Zealand, pp.190–199.

Heisey, P. and Brennan, J.P. (1991) An analytical model of farmers demands for replacements. *American Journal of Agricultural Economics* 73(4), 1044–1052.

Heydecker, W. (1978) Primed seed for better crop establishment. *Span* 21(1), 12–14.

Heydecker, W. and Gibbins, B.M. (1978) The 'priming' of seeds. *Acta Horticulturae* 83, 213–223.

Holden, J.H.W. (1984) Oats. In: Simmonds, N.W. (ed.) *Evolution of Crop Plants*. Longman, London and New York, pp.86–90.

Hole, C.C. and Hardwick, R.C. (1977) Chemical aids to drying seeds of beans (*Phaseolus vulgaris*) before harvest. *Annals of Applied Biology* 88(3), 421–427.

Hrabovszky, J.P. (1982) Crop Production in developing countries in 2000. In: Feistritzer, W.P. (ed.) *Seeds*. FAO, Rome, pp.29–39.

IBPGR (1984) *Descriptors for Soyabean*. International Board for Plant Genetic Resources, Rome.

IFOAM, ISF and FAO (2004a) *Proceedings of First World Conference on Organic Seed*. FAO, Rome.

IFOAM, ISF and FAO (2004b) *Reports of First World Conference on Organic Seed*. FAO, Rome.

IPGRI (1993) *Diversity for Development. The Strategy of the International Plant Genetic Resources Institute*. International Plant Genetic Resources Institute, Rome.

ISTA (1995a) *Understanding Seed Vigour*. International Seed Testing Association, Zurich.

ISTA (1995b) *Handbook of Vigour Test Methods*, 3rd edn. International Seed Testing Association, Zurich.

ISTA (2007) *ISTA List of Stabilized Plant Names*, 5th edn. International Seed Testing Association, Zurich.

ISTA (2009a) *International Rules for Seed Testing*. International Seed Testing Association, Zurich.

ISTA (2009b) *Handbook on Pure Seed Definitions*, 3rd edn. International Seed Testing Association, Zurich.

Jeffs, K.A. (ed.) (1986) *Seed Treatment*, 2nd edn. The British Crop Protection Council, Thornton Heath, UK, 332 pp.

Jianhua, Z., McDonald, M.B. and Sweeney, P.M. (1996) Soybean cultivar identification using RAPD. *Seed Science and Technology* 24, 589–592.

Jones, L.H. (1963) The effect of soil moisture gradients on the growth and development of broad bean (*Vicia faba* L.). *Horticultural Research* 3, 13–26.

Kelly, A.F. (1978) The role of seed programmes in agricultural development. In: Feistritzer W.P. and Kelly, A.F. (eds) *Improved Seed Production.* FAO, Rome, pp.13–24.

Kelly, A.F. (1994) *Seed Planning and Policy for Agricultural Production, the Roles of Government and Private Enterprise in Supply and Distribution.* Bellhaven Press, London and New York.

Kelly, A.F. and George R.A.T. (1998a) Reserve seed stocks against major disasters. In: *Encyclopaedia of Seed Production of World Crops.* Wiley, Chichester, UK, pp.147–149.

Kelly, A.F. and George R.A.T. (1998b) International agreement and national legislation. In: *Encyclopaedia of Seed Production of World Crops.* Wiley, Chichester, UK, pp.19–67.

Kelly, A.F. and George R.A.T. (1998c) Table 7.1: Isolation of maize, border rows to be discarded to achieve adequate isolation. In: *Encyclopaedia of Seed Production of World Crops.* Wiley, Chichester, UK, pp.239–240.

Khan, A.A. (1992) Preplant physiological seed conditioning. In: Janick, J. (ed.) *Horticultural Reviews,* Vol. 13. John Wiley, New York, pp. 131–181.

Ladizinsky, G. (1999) Identification of lentil's wild genetic stock. *Genetic Resources and Crop Evolution* 46, 115–118.

Langer, R.H.M. and Hill, G.D. (1991) *Agricultural Plants.* Cambridge University Press, Cambridge, UK.

Larter, E.N. (1984) Triticale. In: Simmonds, N.W. (ed.) *Evolution of Crop Plants.* Longman, London and New York, pp.117–120.

Laverack, G.K. and Turner, M.R. (1995) Roguing seed crops for genetic purity: a review. *Plant Varieties and Seeds* 8, 29–46.

Legro, R.J. (2004) Organic seed and coating technology: a challenge and opportunity. In: IFOAM, ISF and FAO (2004) *Proceedings of First World Conference on Organic Seed.* FAO, Rome, pp.108–110.

Lemonius, M. (1998) Asia and Pacific Seed Association (APSA). In: Kelly, A.F. and George, R.A.T. (eds) *Encyclopaedia of Seed Production of World Crops.* John Wiley and Sons, Chichester, UK, pp.39–40.

Lesins, K. (1984) Alfalfa, lucerne. In: Simmonds, N.W. (ed.) *Evolution of Crop Plants.* Longman, London and New York, pp.165–168.

Longden, P.C. (1975) Sugar beet seed pelleting. *ADAS Quarterly Review* 18, 73–80.

Lookhart, G.L. and Wrigley C.W. (1995) Variety identification by electrophoretic analysis. In: Wrigley, C.W. (ed.) *Identification of Food Grain Varieties.* AACC, USA, pp.55–72.

Lutman, P.J.W. (1991) *Weeds in Oilseed Crops.* HGCA Oilseeds Research Review No. OS2. HGCA, London.

Maguire, J.D., Kropf, J.P. and Steen, K.M. (1973) Pea seed viability in relation to bleaching. *Proceedings of the Association of Official Seed Analysts* 63, 51–58.

Martiniello, P. (1992) Morphological aspects involved in varietal registration and the production of certified seed of typical Mediterranean leguminous forage crops. *Plant Varieties and Seeds* 5, 71–82.

Maude, R.B. and Keyworth, W.G. (1967) A new method for the control of seed-borne fungal disease. *Seed Trade Review* 19, 202–204.

McNaughton, I.H. (1984a) Swedes and rapes. In: Simmonds, N.W. (ed.) *Evolution of Crop Plants.* Longman, London and New York, pp.53–56.

McNaughton, I.H. (1984b) Turnip and relatives. In: Simmonds, N.W. (ed.) *Evolution of Crop Plants.* Longman, London and New York, pp.45–48.

Morell, M.K., Peakall, R., Appels, R., Preston, L.R. and Lloyd, H.L. (1995) DNA profiling techniques for plant variety identification. *Australian Journal of Experimental Agriculture* 35, 807–819.

Musopole, E. (1995) Actionaid seed involvement. In: *Proceedings of Workshop on Improved On-farm Seed Production for SADC Countries.* FAO, Rome, pp.47–49.

NIAB (1996) *Oilseeds Variety Handbook.* National Institute of Agricultural Botany, Cambridge, UK, pp.9–40.

Nieuwhof, M. (1969) *Cole Crops.* Leonard Hill, London.

OECD (1995) *OECD Scheme for the Varietal Certification of Sugar Beet and Fodder Beet Moving in International Trade.* OECD, Paris.

Ovalle, C.M., del Pozo, A., Fernandez, F., Chavarria, J. and Arredondo, S. (2010) Arrowleaf clover (*Trifolium vesiculosum* Savi): a new species of annual legumes for high rainfall areas of the Mediterranean Climate Zone of Chile. *Chilean Journal of Agricultural Research* 70(1), 170–177.

Pasumarty, S.V., Higuchi, S. and Murata, T. (1995) Environmental influences on seed yield components of white clover. *Journal of Applied Seed Production* 13, 25–31.

Payne, R.C. (1993a) *Growth Chamber-Greenhouse Testing Procedures: Variety Identification*. International Seed Testing Association, Zurich.

Payne, R.C. (1993b) *Rapid Chemical Identification Techniques*. International Seed Testing Association, Zurich.

Peglar, R.A.D. (1976) Harvest ripeness in grass seed crops. *Grass and Forage Science* 31, 7–13.

Perry, D.A. and Harrison, J.G. (1973) Causes and development of hollow heart in pea seed. *Annals of Applied Biology* 73(1), 95–101.

Peterson, S.S., Baird, C.R., Bitner, R.M. and Idaho, C. (1992) Current status of the Alfalfa leafcutting bee, *Megachile rotundata* as a pollinator of Alfalfa. *Bee Science* 2, 135–142.

PGRO (1978) *Information Sheet Number 70*. Processors' and Growers' Research Organisation. (PGRO), Peterborough, UK.

Pickett, A.A. (1998) Genetic quality. In: Kelly, A.F. and George, R.A.T. (eds) *Encyclopaedia of Seed Production of World Crops*. John Wiley and Sons, Chichester, UK, pp.71–79.

Pritchard, A.J. and Mannetje, L.'t. (1967) The breeding systems and some interspecific relations of a number of African *Trifolium* spp. *Euphytica* 16, 324–329.

Purseglove, J.W. (1974) *Tropical Crops: Monocotyledons*, 5th edn. Longman, London.

Purseglove, J.W. (1984) Millets. In: Simmonds, N.W. (ed.) *Evolution of Crop Plants*. Longman, London and New York, pp. 91–93.

Purseglove, J.W. (1985) *Tropical Crops: Dicolyledons*, 5th edn. Longman, London.

Reisch, W. (1952) Variabilitätsstudien an *Vicia faba* L.Z. *Acker-u PflBau* 94, 281–306.

Richardson, M.J. (1990) *An Annotated list of Seed-borne Diseases*, 4th edn. ISTA, Basserdorf, Switzerland.

Riggs, T.J. (1987) Breeding F1 hybrid varieties of vegetables. In: Feistritzer, W.P. and Kelly, A.F. (eds) *Hybrid Production of Selected Cereal Oil and Vegetable Crops*. FAO, Rome, pp.149–173.

Risi, J.C. and Galway, N.W. (1994) The Chenopodium grains of the Andes, Inca crops for modern agriculture. *Advances in Applied Biology* 10, 145–216.

Roberts, E.H. (1972) Storage environment and the control of viability. In: Roberts, E.H. (ed.) *Viability of Seeds*. Chapman and Hall, London, pp.14–58.

Roberts, E.H. and Roberts, D.L. (1972) Viability nomographs. In: Roberts, E.H. (ed.) *Viability of Seeds*. Chapman and Hall, London, pp.417–123.

Rowse, H.R. (1996a) Drum priming – a non-osmotic method of priming seeds. *Seed Science and Technology* 24, 281–294.

Rowse, H.R. (1996b) Drum priming – an environmentally-friendly way of improving seed performance. *Journal of the Royal Agricultural Society of England* 157, 77–83.

Salter, P.J. (1978a) Techniques and prospects for 'fluid drilling' of vegetable crops. *Acta Horticulturae* 72, 101–108.

Salter, P.J. (1978b) Fluid drilling of pre-germinated seeds: progress and possibilities. *Acta Horticulturae* 83, 245–249.

Salter, P.J. and Goode, J.E. (1967) *Crop Responses to Water at Different Stages of Growth*. Commonwealth Agricultural Bureaux, Farnham Royal, UK.

Sapirstein, H.D. (1995) Variety identification by digital analysis. In: Wrigley, C.W. (ed.) *Identification of Food Grain Varieties*. AACC, USA, pp.91–130.

Sauer, J.D. (1976) Grain amaranths. In: Simmonds, N.W. (ed.) *Evolution of Crop Plants*. Longman, London and New York, pp.4–7.

Schoen, J.F. (1993) *Laboratory Tests for Varietal Determination with Fungal Pathogens*. International Seed Testing Association, Zurich.

Scott, J.M. (1989) Seed coatings and treatments and their effects on plant establishment. *Advances in Agronomy* 42, 43–83.

Seidel, J. (2010) *Impact of Toad Rush on Subterranean Clover Seed Production*. Rural Industries Research and Development Corporation (RIRDC), Kingston, Australia.

Simmonds, N.W. (1984) Quinoa and relatives. In: Simmonds, N.W. (ed.) *Evolution of Crop Plants*. Longman, London and New York, pp.29–30.

Smartt, J. (1976) *Pulses*. Longman, London.

Smith, F.L. (1955) The effects of dates of harvest operations on yield and quality of pink beans. *Hilgardia* 24, 37–52.

Smith, J.S.C. and Smith, O.S. (1991) Restriction fragment length polymorphisms can differentiate among US maize hybrids. *Crop Science* 31, 893–899.

Smith, J.S.C. and Smith, O.S. (1992) Fingerprinting crop varieties. *Advances in Agronomy* 47, 85–140.

Smith, P.M. (1984) Sanfoin. In: Simmonds, N.W. (ed.) *Evolution of Crop Plants*. Longman, London and New York, pp.313–314.

Stanley, R. and Wilcockson, S. (2010) Better Organic Bread: Integrating raw material and process requirements for organic bread production. *Project Report No. 474*, Cereals and Oilseeds Division (HGCA) of the Agriculture and Horticulture Development Board (HGCA), Kenilworth, UK.

Still, D.W. and Bradford, K.G. (1998) Using hydrotime and ABA-time models to quantify seed quality of brassicas during development. *Journal of the American Society for Horticultural Science* 123(4), 692–699.

Takahashi, O. (1987) Utilization and seed production of hybrid vegetable varieties in Japan. In: Feistritzer, W.P. and Kelly, A.F. (eds) *Hybrid Production of Selected Cereal Oil and Vegetable Crops*. FAO, Rome, pp.313–328.

Taylor, A.G., Min, T.G. and Mallaber, C.A. (1991) Seed coating system to upgrade Brassicaceae seed quality by exploiting sinapine leakage. *Seed Science and Technology* 19(2), 423–434.

Thoday, P.R. (2011) *Cultivar history of Cultivated Plants*. Thoday Associates, Box Hill, Corsham, UK.

Thompson, J. (2007) Genetically modified crops – good or bad for Africa? *Biologist* 54(3), 129–133.

Thompson, K.F. (1964) Triple-cross hybrid kale. *Euphytica* 13, 173–179.

Thompson, K.F. (1984) Cabbages, kales etc. In: Simmonds, N.W. (ed.) *Evolution of Crop Plants*. Longman, London and New York, pp.49–59.

Thomson, J.R. (1979) *An Introduction to Seed Technology*. Leonard Hill, Blackie Group, Glasgow, UK.

Tindall, H.D. (1983) *Vegetables in the Tropics*. Macmillan Press, London.

Tohme, J., Orlado Gonzalez, D., Beebe, S. and Duque, M.C. (1996) AFLP analysis of gene pools of a wild bean core collection. *Crop Science* 36, 1375–1384.

Tonkin, J.H.B. (1979) Pelleting and other pre-sowing treatments. *Research and Technology of Seeds* 4, 84–105.

Van der Burg, J. (2009) Raising seed quality: what is in the pipeline? *Proceedings of the World Seed Conference*, Rome, 8–10 September, 2009. International Seed Federation (ISF), 1260 Nyon, Switzerland, pp.177–185.

van der Maesen, L.H.G. (1987) Origin, history and taxonomy of chickpea. In: Saxena, M.C. and Singh, K.B. (eds) *The Chickpea*. CAB International, Wallingford, UK.

van de Vooren, J.G. and van der Heijden, G.W.A.M. (1993) Measuring the size of French beans with image analysis. *Plant Varieties and Seeds* 6, 47–53.

Vannozzi, G.P. (1987) Technical and economic aspects of seed production of hybrid varieties of sunflower. In: Feistritzer, W.P. and Kelly, A.F. (eds) *Hybrid Seed Production of Selected Cereal, Oil and Vegetable Crops*. FAO, Rome, pp.253–280.

Vernon, J. (1999) Vibrant colours of grain amaranths. *Journal of The Royal Horticultural Society* 124, 12–15.

Vos, P., Hogers, R., Bleeker, M., van de Lee, T., Hornes, M., Frijters, A., Pot, J., Peleman, J., Kuiper, M. and Zabeau, M. (1995) AFLP: a new technique for DNA fingerprinting. *Nucleic Acids Research* 23, 4407–4414.

Walker, T.S. (2006) *Participatory Varietal Selection, Participatory Plant Breeding and Varietal Change*. World Bank, Washington, DC.

Ward, J.T., Basford, J.H., Hawkins, J.H. and Holliday, M.H. (1985) *Oilseed Rape*. Farming Press, Ipswich, UK.

Wang, Y.R. and Hampton, J.G. (1991) Seed vigour and storage of 'Grasslands Pawera' red clover. *Plant Varieties and Seeds* 4, 61–66.

Weiss, E.A. (1983) *Oilseed Crops*. Longman, London.

Wien, H.C. and Wurr, D.C.E. (1997) Cauliflower, broccoli, cabbage and Brussels sprouts. In: Wien, H.C. (ed.) *The Physiology of Vegetable Crops*. CAB International, Wallingford, UK, pp.511–552.

Williams, J.T. and Brenner D. (1995) Grain amaranth (*Amaranthus* spp.). In: Williams, J.T. (ed.) *Cereals and Pseudocereals*. Chapman and Hall, New York, pp.129–186.

Wilson, A. (1995) An NGO's experience in small scale seed production. In: *Proceedings of Workshop on Improved On-farm Seed Production for SADC Countries*. FAO, Rome, pp.42–46.

Wilson, D.O. and McDonald, M.B. (1992) Mechanical damage in bean (*Phaseolus vulgaris* L.) seed in mechanized and non-mechanized threshing systems. *Seed Science and Technology* 20(3), 571–582.

Wurster, D.E. (1959) Air-suspension technique of coating drug particles. *Journal of the American Pharmaceutical Association* XLVIII, 451–454.

www.aatf-africa.org (2011) African Agricultural Technology Foundation, Project 4: Water Efficient Maize for Africa (WEMA). Accessed 4 February, 2011.

General Index

Page numbers in **bold** refer to illustrations and tables.

Index of Species